Modern Methods in Equilibrium Statistical Mechanics

SERIES ON ADVANCES IN STATISTICAL MECHANICS

Editor-in-Chief: M. Rasetti

Vol. 1: **Integrable Systems in Statistical Mechanics**
Editors: G. M. D'Ariano, A. Montorsi & M. Rasetti

Series on Advances in Statistical Mechanics – Volume 2

MODERN METHODS IN EQUILIBRIUM STATISTICAL MECHANICS

MARIO RASETTI

World Scientific

Published by

World Scientific Publishing Co Pte Ltd.
P. O. Box 128, Farrer Road, Singapore 9128.
242, Cherry Street, Philadelphia PA 19106-1906, USA

Library of Congress Cataloging-in-Publication Data

Rasetti, Mario.
Modern methods in equilibrium statistical mechanics.

1. Statistical mechanics. I. Title.
QC174.8.R37 1986 530.1'3 85-29637
ISBN 9971-966-27-1
ISBN 9971-966-29-8 (pbk.)

Printed in Singapore by Kyodo-Shing Loong Printing Industries Pte Ltd.

To Tiziana

Foreword

This book arose out of a set of graduate courses in Analytical Mechanics and Statistical Mechanics which I successively gave at the University of Miami, in Coral Gables, Florida and at the University and Polytechnic of Turin, Italy over the past decade.

It is intended for students in physics, mathematics and engineering, both graduate and undergraduate, who are interested in approaching the tremendously rich, and yet in large part still unexplored, domain of nonlinear phenomena in mechanics and statistical physics.

As for mechanics, only the features of randomness and chaoticity induced by nonlinearities are considered, whereas the equally fascinating realm of deterministic dynamics (e.g., soliton solutions) is regretfully left out. Indeed, Statistical Mechanics was the final aim of these lectures, and the whole book is centered on the idea that stochasticity can come out of nonlinearities even in the case of a few degrees of freedom, and on how this bears on the known methods of classical statistical mechanics and its link with thermodynamics.

For these reasons the final part of the book is a review of the mathematical tools which allow us to bridge, in a consistent and elegant manner, the two ways leading to randomness: fluctuations, on the one hand, connected with the essentially infinite number of degrees of freedom of any real physical system, and chaos, on the other, mainly due to the topological complexity of orbits in phase space even for very simple nonlinear dynamical systems. To this latter aspect is devoted the first part of and main bulk of the book.

It should be emphasized that the book does not intend in any way to be complete or exhaustive. The subject is in constant and active progress and a wealth of results both numerical and analytical is accumulating day after day.

I would like to acknowledge here my dear and unforgettable friend and master, Lars Onsager, who first introduced me to this fascinating subject and to its subtle beauty; and to thank, in particular, my friends, V. de Alfaro and T. Regge, for helping me to understand it. I hope that my innumerable other colleagues to whom I owe my comprehension of several specific points will forgive me for not mentioning them explicitly here.

Mario Rasetti

Turin
December 1985

Contents

Chapter 1
Review of Basic Classical Mechanics

1. Newtonian Mechanics of a System of Particles

The dynamics of a single particle of mass m (assumed to be constant, no relativistic effects being considered) whose position in space (unless otherwise specified $\mathbb{R}^3$ or a finite part Ω thereof) is defined by the position vector $\vec{q} = \vec{q}(t)$ at time t, is completely described by the equation of motion (Newton's second law)

$$\frac{d}{dt}(m\dot{\vec{q}}) = \frac{d}{dt}(m\vec{v}) = \vec{F} \quad , \tag{1.1}$$

in which $\vec{F}$ is the total force acting on the particle and $\vec{v}$ its velocity. A dot is used when no ambiguity arises to denote a time derivative. It is interesting to note that the formulation (1.1) — which Newton himself used — remains valid even in the case of variable m, which is not the case for the more familiar form of the equation of motion (1.1),

$$\vec{F} = m\vec{a} = m\ddot{\vec{q}} \quad , \tag{1.2}$$

where $\vec{a}$ is the acceleration of the particle.

If the particle belongs to an isolated system, consisting of N mass points, Eq. (1.1) (or (1.2)) still holds for each particle and the dynamics of the system is described by the system of N equations

$$m_i \ddot{\vec{q}}_i = \vec{F}_i = \sum_{\substack{j=1 \\ j \neq i}}^{N} \vec{F}_{ij} \quad , \quad i = 1,\ldots,N \quad . \tag{1.3}$$

Here m_i, $\vec{q}_i$ $(i = 1,\ldots,N)$ denote, respectively, the mass and position of the i-th particle in the system. Since the system has been assumed to be isolated, namely the only forces acting on the particles in the system are mutual inter-particle forces, the force $\vec{F}_i$ acting on particle i has been written in (1.3) as the vector sum of the (N-1) forces $\vec{F}_{ij}$ exerted by all other particles j $(j \neq i)$ on particle i.

Particularly interesting is the case when the forces $\vec{F}_{ij}$ are conservative, i.e. can be derived from a two-body potential U_{ij}:

$$\vec{F}_{ij} = - \nabla_{\vec{q}_i} U_{ij} \quad , \tag{1.4}$$

where $\nabla_{\vec{q}_i}$ is the gradient operator with respect to the coordinates of particle i. In most cases, U_{ij}, which describes the potential energy of the pair, is function only of the distance between the two particles i,j

$$U_{ij} = U_{ij}(|\vec{q}_i - \vec{q}_j|) \doteq U_{ij}(r_{ij}) = U_{ji} \quad . \tag{1.5}$$

In this case Newton's third law, usually referred to as the action-reaction principle, holds

$$\vec{F}_{ij} = - \frac{dU_{ij}}{dr_{ij}} \nabla_{\vec{q}_i} r_{ij} = - \frac{dU_{ij}}{dr_{ij}} \frac{\vec{q}_i - \vec{q}_j}{r_{ij}} \equiv \nabla_{\vec{q}_j} U_{ji} = - \vec{F}_{ji} \, , \tag{1.6}$$

namely the force exerted on particle i by particle j is opposite to the force due to particle i upon particle j; moreover

$$\vec{F}_i = - \nabla_{\vec{q}_i} U \tag{1.7}$$

with

$$U = \frac{1}{2} \sum_{\substack{i,j=1 \\ i \neq j}}^{N} U_{ij}(r_{ij}) \quad . \tag{1.8}$$

Equation (1.6) contains two more pieces of information. Under the assumption (1.5) the mutual force between two generic particles i,j in the system is central (i.e. acting along the line connecting the two bodies and of the form $\vec{F}_{ij} = f_{ij}(r_{ij})\vec{u}_{ij}$, where $f_{ij} = f_{ji}$ is a scalar function of the mutual distance r_{ij}, and $\vec{u}_{ij} = (\vec{q}_i - \vec{q}_j)/r_{ij} = -\vec{u}_{ji}$ the unit vector along such line). Moreover the total momentum of the system is conserved. Indeed, upon summing both sides of (1.3) with respect to i, one gets

$$\sum_{i=1}^{N} \vec{F}_i = \sum_{\substack{i,j=1 \\ i \neq j}}^{N} \vec{F}_{ij} = \sum_{i=1}^{N} m_i \ddot{\vec{q}}_i = 0 \quad . \tag{1.9}$$

The latter, upon time integration, implies

$$\sum_{i=1}^{N} m_i \dot{\vec{q}}_i \doteq \vec{P} = \text{const.} \tag{1.10}$$

$m_i \dot{\vec{q}}_i$ is but the i-th particle momentum (see (1.1)) and the conserved quantity $\vec{P}(\dot{\vec{P}} = 0)$ of (1.10) is, by definition, the total momentum of the system. Also from (1.3) one has

$$\sum_{i=1}^{N} \vec{q}_i \times \vec{F}_i = \sum_{i=1}^{N} \vec{q}_i \times \frac{d}{dt}(m_i \dot{\vec{q}}_i) =$$

$$= \frac{d}{dt}\left[\sum_{i=1}^{N} (\vec{q}_i \times m_i \dot{\vec{q}}_i) \right] \quad . \tag{1.11}$$

$\vec{q}_i \times m_i \dot{\vec{q}}_i$ is the i-th particle angular momentum, and

$$\vec{L} = \sum_{i=1}^{N} (\vec{q}_i \times m_i \dot{\vec{q}}_i) \quad , \tag{1.12}$$

the total angular momentum of the system, whereas

$$\vec{M} \doteq \sum_{T=1}^{N} \vec{q}_i \times \vec{F}_i \qquad (1.13)$$

is the torque (the total moment) exerted by the forces acting in the system. One can write, by (1.6),

$$\begin{aligned}\vec{M} = \sum_{\substack{i,j=1\\ i\neq j}}^{N} \vec{q}_i \times \vec{F}_{ij} &= \sum_{i,j=1}^{N} (\vec{q}_i - \vec{q}_j) \times \vec{F}_{ij} + \sum_{\substack{i,j=1\\ i\neq j}}^{N} \vec{q}_j \times \vec{F}_{ij} = \\ &= - \sum_{\substack{i,j=1\\ i\neq j}}^{N} \vec{q}_j \times \vec{F}_{ji} = - \vec{M} \qquad (1.14)\end{aligned}$$

where the property that $\vec{F}_{ij}$ is central (that is, $(\vec{q}_i - \vec{q}_j) \times \vec{F}_{ij} = 0$) has been used. Equation (1.14) obviously implies $\vec{M} = 0$, and (1.11) reads

$$\dot{\vec{L}} = 0 \quad , \qquad \vec{L} = \text{const.} \qquad (1.15)$$

In other words the total angular momentum of the system is also a conserved quantity.

Notice that the conservation of linear momentum $\vec{P}$ depends on the invariance of the system of forces $\vec{F}_{ij}$ with respect to translations (i.e. transformations of the configuration space of the system $\mathbb{R}^{3N}$ of the form $\vec{q}_i \to \vec{q}_i + \vec{a}$, $i = 1,\dots,N$ where $\vec{a} \in \mathbb{R}^3$ is a constant vector) expressed by (1.5). Analogously the conservation of angular momentum $\vec{L}$ depends on the invariance of the system of moments $\vec{q}_i \times \vec{F}_{ij}$ with respect to rotations around an axis through the origin of direction $\vec{u}$ (an infinitesimal transformation of this type is $\vec{q}_i \to \vec{q}_i + \delta\theta \vec{u} \times \vec{q}_i$, $i = 1,\dots,N$) exhibited by (1.14). These are special cases of a much more general property of dynamical systems — described by Noether's theorem — whereby, for every group of transformations of the configuration space leaving the system invariant (in a sense which will be made more precise in the sequel), there exists a corresponding constant of the motion.

The equations of motion (1.3) can on the other hand be derived, in a quite general way, from a variational principle. The latter states that the true orbit $\vec{q}_i(t)$, $i = 1,\dots,N$ of the system in configuration space, i.e., the solution of (1.3) for a specified set of initial conditions, is that which makes the action integral

$$A = \int_{t_1}^{t_2} \mathscr{L}\, dt \quad , \tag{1.16}$$

stationary (indeed minimum). Here $\mathscr{L}$ is the Lagrange function of the system, describing the difference between the total kinetic and potential energies.

$$\mathscr{L} = \mathscr{L}(\{\vec{q}_i, \dot{\vec{q}}\}) = \sum_{i=1}^{N} \frac{1}{2} m_i \dot{\vec{q}}_i^{\,2} - U \quad . \tag{1.17}$$

The integral in (1.16) is a functional of the orbit, i.e. it is to be thought of as calculated along a given trajectory $\vec{q}_i = \vec{q}_i(t)$, $i = 1,\dots,N$, such that at time $t = t_1$, $\vec{q}_i = \vec{q}_i^{(1)}$ and at $t = t_2$, $\vec{q}_i = \vec{q}_i^{(2)}$ with $\vec{q}_i^{(1)}$, $\vec{q}_i^{(2)}$ $(i = 1,\dots,N)$ denoting fixed points in configuration space. Indeed, upon performing arbitrary, small variations $\delta\vec{q}_i(t)$ on $\vec{q}_i(t)$

$$\vec{q}_i(t) \to \vec{q}_i(t) + \delta\vec{q}_i(t) \quad , \quad i = 1,\dots,N \tag{1.18}$$

such that

$$\delta\vec{q}_i(t_1) = \delta\vec{q}_i(t_2) = 0 \quad , \quad i = 1,\dots,N \tag{1.19}$$

(in other words considering two neighbouring orbits which at times t_1 and t_2 pass through the same points $\vec{q}_i^{(1)}$ and $\vec{q}_i^{(2)}$ respectively, $\forall i$), to the first order in the $\delta\vec{q}_i$'s the action varies by

$$\delta A = \int_{t_1}^{t_2} \sum_{j=1}^{N} \{\nabla_{\vec{q}_j} \mathscr{L} \cdot \delta\vec{q}_j + \nabla_{\dot{\vec{q}}_j} \mathscr{L} \cdot \delta\dot{\vec{q}}_j\} \, dt =$$

$$= \int_{t_1}^{t_2} \sum_{j=1}^{N} \{m_j \dot{\vec{q}}_j \cdot \delta\dot{\vec{q}}_j - \sum_{\substack{k=1 \\ k\neq j}}^{N} \nabla_{\vec{q}_j} U_{jk} \cdot \delta\vec{q}_j\} \, dt \quad . \qquad (1.20)$$

Upon integrating by parts, taking into account the identities $\delta\dot{\vec{q}}_i \doteq \frac{d}{dt}(\vec{q}_i + \delta\vec{q}_i) - \frac{d}{dt}\vec{q}_i \equiv \frac{d}{dt}\delta\vec{q}_i \, , \forall i,$ as well as the boundary conditions, (1.19), (1.20) can be cast in the form

$$\delta A = \int_{t_1}^{t_2} \sum_{j=1}^{N} \left\{ -m_j \ddot{\vec{q}}_j - \sum_{\substack{k=1 \\ k\neq j}}^{N} \nabla_{\vec{q}_j} U_{jk} \right\} \cdot \delta\vec{q}_j(t) \, dt \quad . \qquad (1.21)$$

Since the $\delta\vec{q}_i(t)$'s for $t \neq t_1, t_2$ are arbitrary functions of time, $\delta A = 0$ is realized only by those trajectories which make the factors in curly bracket in (1.21) vanish, namely just the solutions of Newton's equations of motion (1.3). This proves the equivalence of the latter with the requirement of stationarity for the action integral.

2. Hamilton - Lagrange Equations of Motion

The variational principle of the previous section allows us to write the equations of motion in a form which does not resort explicitly to the forces acting within the system but only to the Lagrange function. Indeed if integration by parts is performed in (1.20) before introducing for $\mathscr{L}$ the explicit form (1.17), the requirement of stationarity $\delta A = 0$ leads to

$$\frac{d}{dt}\frac{\partial \mathscr{L}}{\partial \dot{q}_\alpha} - \frac{\partial \mathscr{L}}{\partial q_\alpha} = 0 \quad , \quad \alpha = 1,\ldots,3N \qquad (2.1)$$

(where the degrees of freedom have been relabelled by their Cartesian components).

Equation (2.1) is referred to as the Lagrangian formulation of the equations of motion. One recovers these (indeed in the form (1.1)) by introducing the generalized momenta

$$p_\alpha \doteq \frac{\partial \mathscr{L}}{\partial \dot{q}_\alpha} \quad , \qquad \alpha = 1,\ldots,3N \quad , \tag{2.2}$$

and the generalized forces

$$F_\alpha \doteq \frac{\partial \mathscr{L}}{\partial q_\alpha} \quad , \qquad \alpha = 1,\ldots,3N \quad . \tag{2.3}$$

The advantage of (2.1) is that it is independent on the choice of the coordinates with which one describes the system, and indeed the q_α, $\alpha = 1,\ldots,3N$ could just as well be arbitrary differentiable functions of the effective configuration space Cartesian positions (we denote them now by $\vec{x}_i$, $i = 1,\ldots,N$). The $\dot{q}_\alpha$, $\alpha = 1,\ldots,3N$ in this case would be generalized velocities, connected to the true velocities $\dot{\vec{x}}_i \doteq \vec{v}_i$, $i = 1,\ldots,N$ by

$$\dot{\vec{x}}_i = \sum_{\alpha=1}^{3N} \frac{\partial \vec{x}_i}{\partial q_\alpha} \dot{q}_\alpha \quad . \tag{2.4}$$

Upon setting, in general

$$\mathscr{L}(\{q_\alpha, \dot{q}_\alpha\}) = K(\{q_\alpha, \dot{q}_\alpha\}) - V(\{q_\alpha\}) \quad , \tag{2.5}$$

where K denotes the total kinetic energy, the potential energy V — being independent of the generalized velocities — is simply obtained from $U(\{\dot{x}_i\})$ by the coordinate transformation $\{\vec{x}_i, i = 1,\ldots,N\} \rightarrow \{q_\alpha; \alpha = 1,\ldots,3N\}$. Since the transformation (2.4) is linear, K turns out immediately to be a homogeneous quadratic function in the $\dot{q}_\alpha$'s;

$$K = K(\{q_\alpha,\dot{q}_\alpha\}) = \frac{1}{2} \sum_{\alpha,\beta=1}^{3N} a_{\alpha\beta}\, \dot{q}_\alpha\, \dot{q}_\beta \quad , \tag{2.6}$$

where

$$a_{\alpha\beta} = a_{\beta\alpha} \doteq \sum_{i=1}^{N} m_i \frac{\partial \vec{x}_i}{\partial q_\alpha} \cdot \frac{\partial \vec{x}_i}{\partial q_\beta} \quad . \tag{2.7}$$

Using (2.5) with (2.6) in (2.1), one gets the equations of motion in the form

$$\sum_{\beta=1}^{3N} a_{\alpha\beta}\, \ddot{q}_\beta + \sum_{\beta,\gamma=1}^{3N} \left\{ \left(\frac{\partial a_{\alpha\beta}}{\partial q_\gamma} - \frac{1}{2} \frac{\partial a_{\beta\gamma}}{\partial q_\alpha} \right) \dot{q}_\beta\, \dot{q}_\gamma \right\} + \frac{\partial V}{\partial q_\alpha} = 0 \quad ,$$

$$\alpha = 1,\ldots,3N \quad . \tag{2.8}$$

If one defines the Christoffel symbol

$$\begin{Bmatrix} \beta\gamma \\ \alpha \end{Bmatrix} \doteq \frac{1}{2} \left(\frac{\partial a_{\alpha\beta}}{\partial q_\gamma} + \frac{\partial a_{\gamma\alpha}}{\partial q_\beta} - \frac{\partial a_{\beta\gamma}}{\partial q_\alpha} \right) \tag{2.9}$$

one may write (2.8) in the form

$$\sum_{\beta=1}^{3N} a_{\alpha\beta}\, \ddot{q}_\beta = F_\alpha - \sum_{\beta,\gamma=1}^{3N} \begin{Bmatrix} \beta\gamma \\ \alpha \end{Bmatrix} \dot{q}_\beta\, \dot{q}_\gamma \doteq \mathscr{F}_\alpha \quad , \quad \alpha = 1,\ldots,3N \tag{2.10}$$

which is a generalized formulation of Newton's equation with tensor, coordinate dependent mass a, and coordinate and velocity dependent force $\mathscr{F}$.

Naturally in the case when, as already suggested, the q_α are simply the components of the $\vec{x}_i$'s, one recovers exactly the equation (1.17). In order to check this, set, e.g., $(\vec{x}_i)_\xi = q_{3(i-1)+\xi}$; $i = 1,\ldots,N$ ($\xi = 1,2,3$ labelling the Cartesian components). Then a is diagonal

$$a_{\alpha\beta} = m_{[\![\frac{\alpha-1}{3}]\!] + 1} \, \delta_{\alpha\beta} \tag{2.11}$$

($\delta_{\alpha\beta}$ being the Kronecker symbol, and $[\![x]\!]$, $x \in \mathbb{R}$ denoting the maximum integer $\leq x$). Moreover $\{{}^{\beta\alpha}_{\alpha}\} \equiv 0$, and $\mathscr{F}$ is the true force

$$F_\alpha = - \frac{\partial V}{\partial q_\alpha} = - \sum_{i=1}^{N} \nabla_{\vec{x}_i} U \cdot \frac{\partial \vec{x}_i}{\partial q_\alpha} = \left(\vec{F}_{[\![\frac{\alpha-1}{3}]\!] + 1} \right)_{\alpha - 3[\![\frac{\alpha-1}{3}]\!]} \tag{2.12}$$

namely $(\vec{F}_i)_\xi = F_{3(i-1)+\xi}$.

One can therefore now identify the configuration space with a manifold M of dimension $n = 3N$ with coordinates $\vec{Q} \equiv \{q_\alpha;\ \alpha = 1,\ldots,n\}$, and write (2.1) in the form

$$\frac{d}{dt} \nabla_{\dot{\vec{Q}}} \mathscr{L} - \nabla_{\vec{Q}} \mathscr{L} = 0 \quad . \tag{2.13}$$

One says that a regular transformation of M into itself, $h : M \to M$ leaves the dynamical system invariant if the Lagrange function $\mathscr{L}(\vec{Q}, \dot{\vec{Q}})$ (which is a smooth real valued function over the tangent bundle TM of M) is such that for any tangent vector $\vec{v} \in TM$

$$\mathscr{L}(h \circ \vec{Q}, h_* \vec{v}) = \mathscr{L}(\vec{Q}, \vec{v}) \quad . \tag{2.14}$$

Then Noether's theorem states: if there exists a one-parameter group $\mathscr{h}$ of transformations h_s, $s \in \mathbb{R}$ such that h_0 = identity, leaving the dynamical system invariant for all s, the equations of motion (2.13) admit a constant of motion of the form

$$I = I(\vec{Q}, \dot{\vec{Q}}) = \nabla_{\dot{\vec{Q}}} \mathscr{L} \cdot \left. \frac{dh_s(\vec{Q})}{ds} \right|_{s=0} \quad . \tag{2.15}$$

In order to prove (2.15), let $\vec{Q}_0 = \vec{Q}_0(t)$ be a solution of Lagrange's equation (2.13). Since, by hypothesis, the dynamical system is left invariant by h, for any arbitrary s, $h_s \circ \vec{Q}_0(t) \doteq \vec{Q}_s(t)$ is still a solution of Lagrange's equations, namely,

$$\frac{\partial}{\partial t}\left[\nabla_{\dot{\vec{Q}}_s} \mathscr{L}(\vec{Q}_s(t), \dot{\vec{Q}}_s(t))\right] = \nabla_{\vec{Q}_s} \mathscr{L}(\vec{Q}_s(t), \dot{\vec{Q}}_s(t)) \quad . \qquad (2.16)$$

Moreover, by the assumed invariance of $\mathscr{L}$ under h_s

$$\frac{\partial}{\partial s} \mathscr{L}(\vec{Q}_s, \dot{\vec{Q}}_s) = \nabla_{\dot{\vec{Q}}_s} \mathscr{L} \cdot \frac{\partial \dot{\vec{Q}}_s}{\partial s} + \nabla_{\vec{Q}_s} \mathscr{L} \cdot \frac{\partial \vec{Q}_s}{\partial s} = 0 \quad . \qquad (2.17)$$

Replacing $\nabla_{\vec{Q}_s} \mathscr{L}$ from (2.16) in the right-hand side of (2.17), the latter becomes

$$\nabla_{\dot{\vec{Q}}_s} \mathscr{L} \cdot \frac{\partial}{\partial t}\frac{\partial \vec{Q}_s}{\partial s} + \frac{\partial}{\partial t} \nabla_{\dot{\vec{Q}}_s} \mathscr{L} \cdot \frac{\partial \vec{Q}_s}{\partial s} = 0 =$$

$$= \frac{\partial}{\partial t}\left(\nabla_{\dot{\vec{Q}}_s} \mathscr{L} \cdot \frac{\partial \vec{Q}_s}{\partial s}\right) \quad , \qquad (2.18)$$

thus proving (2.15) (recall that at $s = 0$, h_s is the identity transformation of $\mathscr{h}$).

The equations of motion (1.3) for the particle system

$$m_i \ddot{\vec{q}}_i = \vec{F}_i \quad , \qquad i = 1,\ldots,N \qquad (2.19)$$

constitute a system of N second order differential equations in the variables $\vec{q}_i = \vec{q}_i(t)$, $i = 1,\ldots,N$. The theory of differential equations says that they can always be transformed into a system of 2N first order differential equations, by introducing an auxiliary set of N new dynamical variables $\vec{p}_i = \vec{p}_i(t)$, $i = 1,\ldots,N$. The simplest way to achieve this is to set

$$\vec{p}_i = m_i \dot{\vec{q}}_i \quad , \qquad i = 1,\ldots,N \tag{2.20-a}$$

whereby

$$\dot{\vec{p}}_i = \vec{F}_i \quad . \tag{2.20-b}$$

Notice that, due to (1.17), (2.20-a) is consistent with (2.2) and identifies the $\vec{p}_i$'s with the single particle momenta of the particles,

$$\vec{p}_i = \nabla_{\dot{\vec{q}}_i} \mathscr{L} \quad , \tag{2.21}$$

which, as (2.20-b) is just a restatement of (1.7), is obviously consistent with (2.3).

Taking the total time derivative of $\mathscr{L}$, as given by (1.17), one gets

$$\frac{d\mathscr{L}}{dt} = \frac{\partial \mathscr{L}}{\partial t} + \sum_{i=1}^{N} \nabla_{\dot{\vec{q}}_i} \mathscr{L} \cdot \ddot{\vec{q}}_i + \sum_{i=1}^{N} \nabla_{\vec{q}_i} \mathscr{L} \cdot \dot{\vec{q}}_i \quad . \tag{2.22}$$

Entering in (2.22) the equations of motion, written in the form

$$\vec{F}_i = - \sum_{\substack{j=1 \\ j\neq 1}}^{N} \nabla_{\vec{q}_i} U_{ij} = \nabla_{\vec{q}_i} \mathscr{L} = \frac{d}{dt} \nabla_{\dot{\vec{q}}_i} \mathscr{L} = m_i \ddot{\vec{q}}_i \quad , \quad i = 1,\ldots,N \tag{2.23}$$

it becomes

$$\frac{d\mathscr{L}}{dt} = \frac{\partial \mathscr{L}}{\partial t} + \sum_{i=1}^{N} \nabla_{\dot{\vec{q}}_i} \mathscr{L} \cdot \ddot{\vec{q}}_i + \sum_{i=1}^{N} \left(\frac{d}{dt} \nabla_{\dot{\vec{q}}_i} \mathscr{L} \right) \cdot \dot{\vec{q}}_i =$$

$$= \frac{\partial \mathscr{L}}{\partial t} + \sum_{i=1}^{N} \frac{d}{dt} \left(\nabla_{\dot{\vec{q}}_i} \mathscr{L} \cdot \dot{\vec{q}}_i \right) \quad . \tag{2.24}$$

In other words

$$\frac{d}{dt}\left\{ \sum_{i=1}^{N} \nabla_{\dot{\vec{q}}_i} \mathscr{L} \cdot \dot{\vec{q}}_i - \mathscr{L} \right\} = - \frac{\partial \mathscr{L}}{\partial t} \quad . \tag{2.25}$$

Using (2.21), and noticing (from (1.17)) that $\mathscr{L}$ does not depend explicitly on time, whence $\frac{\partial \mathscr{L}}{\partial t} = 0$; (2.25) implies that the function of dynamical variables

$$H = H(\{\vec{q}_i , \vec{p}_i\}) = \sum_{i=1}^{N} \vec{p}_i \cdot \dot{\vec{q}}_i - \mathscr{L} = \text{const.} \tag{2.26}$$

is a constant of the motion. H is called Hamilton's function (or Hamiltonian) of the system, and by (2.20-a), (1.17) represents the total energy of the system,

$$H = \sum_{i=1}^{N} \frac{\vec{p}_i^{\,2}}{2m_i} + \frac{1}{2} \sum_{\substack{i,j=1 \\ i \neq j}}^{N} U_{ij}(|\vec{q}_i - \vec{q}_j|) \quad , \tag{2.27}$$

sum of the total kinetics and potential energies. One can now repeat the whole variational calculation with

$$\mathscr{L} = \sum_{i=1}^{N} \vec{p}_i \cdot \dot{\vec{q}}_i - H \tag{2.28}$$

namely

$$A = \int_{t_1}^{t_2} \left\{ \sum_{i=1}^{N} \vec{p}_i \cdot \dot{\vec{q}}_i - H(\{\vec{q}_i , \vec{p}_i\}) \right\} dt \tag{2.29}$$

treating the $\vec{p}_i$ and $\vec{q}_i$'s as functionally independent variables. The resulting equations of motion,

$$\dot{\vec{q}}_i = \nabla_{\vec{p}_i} H \quad ,$$

$$\dot{\vec{p}}_i = - \nabla_{\vec{q}_i} H \quad , \qquad i = 1,\dots,N \tag{2.30}$$

are, of course, equivalent to (2.20) by (2.27). Equations (2.30) are known as canonical equations of motion. In canonical language, a dynamical state of the system is thus completely characterized at each time t by the set of vectors $\{\vec{q}_i(t), \vec{p}_i(t); i = 1,\dots,N\}$ which are solutions of (2.30) for a specified set of initial conditions. Such a set, when interpreted as defining a single point in a 6·N-dimensional space Φ (e.g. $\underbrace{\Omega \times \dots \times \Omega}_{N \text{ factors}} \times \mathbb{R}^{3N}$, $\Omega \subset \mathbb{R}^3$ denoting the spatial configuration domain), introduces the concept of phase space. The state representative point describes an orbit in Φ as t varies. If H is the only constant of motion, such an orbit is entirely embedded in a $(6N-1)$-dimensional submanifold Σ of Φ, defined by

$$H(\{\vec{q}_i, \vec{p}_i\}) = \mathscr{E} = \sum_{i=1}^{N} \frac{|\vec{p}_i^{(0)}|^2}{2m_i} + \frac{1}{2} \sum_{\substack{i,j=1 \\ i \neq j}}^{N} U_{ij}(|\vec{q}_i^{(0)} - \vec{q}_j^{(0)}|) \tag{2.31}$$

where $\vec{p}_i^{(0)}, \vec{q}_i^{(0)}$, $i = 1,\dots,N$ are the initial conditions (i.e., the given canonical coordinates at time $t = 0$), and the constant $\mathscr{E}$ is the conserved total energy of the system. If more independent dynamical quantities are conserved — due to specific symmetries of the system — the motion is restricted to lower-dimensional manifolds, intersections of Σ with the submanifolds associated to each constant of the motion in Φ. The equation of motion for any dynamical variable, i.e. any function of the canonical coordinates and of time $F = F(\{\vec{q}_i, \vec{p}_i\}, t)$, is given by

$$\dot{F} = \frac{\partial F}{\partial t} + \sum_{i=1}^{N} (\nabla_{\vec{q}_i} F \cdot \dot{\vec{q}}_i + \nabla_{\vec{p}_i} F \cdot \dot{\vec{p}}_i) =$$

$$= \frac{\partial F}{\partial t} + \sum_{i=1}^{N} (\nabla_{\vec{q}_i} F \cdot \nabla_{\vec{p}_i} H - \nabla_{\vec{p}_i} F \cdot \nabla_{\vec{q}_i} H) \quad , \tag{2.32}$$

where (2.30) have been used.

Upon introducing the notation

$$\{f, g\} = \sum_{i=1}^{N} (\nabla_{\vec{q}_i} f \cdot \nabla_{\vec{p}_i} g - \nabla_{\vec{p}_i} f \cdot \nabla_{\vec{q}_i} g) \quad , \tag{2.33}$$

where f, g denote arbitrary, differentiable functions of $\vec{q}_i, \vec{p}_i$, $i = 1,\ldots,N$, (2.32) can be written

$$\dot{F} = \frac{\partial F}{\partial t} + \{F, H\} \quad . \tag{2.34}$$

$\{f, g\}$ is called the Poisson bracket of f and g. It is straightforward to check that

$$\{f, g\} = -\{g, f\} \quad , \tag{2.35}$$

moreover $\{f, f\} = 0$. We also have

$$\{\vec{q}_i, \vec{q}_j\} = \{\vec{p}_i, \vec{p}_j\} = 0 \quad , \tag{2.36}$$

$$\{\vec{q}_i, \vec{p}_j\} = \delta_{ij} \quad , \tag{2.37}$$

hence

$$\{\vec{p}_i, f\} = -\nabla_{\vec{q}_i} f \quad , \quad \{\vec{q}_i, f\} = \nabla_{\vec{p}_i} f \quad . \tag{2.38}$$

Finally

$$\{f, gh\} = g\{f, h\} + \{f, g\}h \quad , \tag{2.39}$$

and

$$\{f, \{g, h\}\} + \{g, \{h, f\}\} + \{h, \{f, g\}\} = 0 \quad . \tag{2.40}$$

Equation (2.40) is referred to as the Jacobi identity. Equations (2.34), (2.35), (2.39) and (2.40) show that the Poisson bracket transforms the infinite dimensional linear space of differentiable vector fields $\mathscr{I}$ over the manifold Φ (more precisely Σ) into a Lie algebra.

Recall that, by definition, a Lie algebra is a linear space L endowed with an antisymmetric, bilinear operation (customarily denoted by square brackets, instead of curly brackets, and called commutator) $L \times L \to L$ satisfying Jacobi's identity.

We have the following property (which is somewhat a restatement of Noether's theorem). If the variable F does not depend explicitly on time, i.e. $\frac{\partial F}{\partial t} = 0$, (2.34) reads

$$\dot{F} = \{F , H\} \quad . \tag{2.41}$$

Hence F is a constant of the motion if and only if its Poisson bracket with H is zero. The property $\{f , f\} = 0$, then straightforwardly implies that H itself is a constant of the motion.

Equations (2.36) and (2.37) show that Φ is endowed with a symplectic structure; the components $(\vec{q}_i)_\xi , (\vec{p}_i)_\xi$, $i = 1,\dots,N$; $\xi = 1,2,3$ form a complete symplectic basis over Φ. Let us remind ourselves that a Euclidean structure in a linear space is given by a symmetric bilinear form whereas a symplectic structure is given by an antisymmetric form: thus even though very similar in several properties, symplectic geometry is not Euclidean.

3. Integrable Dynamical Systems

The system of equations (2.30) is said to be integrable (and by extension the corresponding dynamical system is likewise integrable) if there exist 3N mutually independent differentiable analytic functions of $\vec{q}_i , \vec{p}_i$, $i = 1,\dots,N$ which are constants of the motion. Notice that the requirement is global, namely, the conserved quantities must exist for all allowed values of the variables. The adjective "integrable" comes from the fact that — in principle — for such a case one could solve the constants of motion to get the $\vec{p}_i$'s (i.e., the $\dot{\vec{q}}_i$'s); whence the evaluation of $\vec{q}_i(t)$, $i = 1,\ldots,N$ would be reducible to quadratures. The existence of 3N constants of motion confines the orbit in Φ to a hyper-surface σ of dimension 3N.

Integrable systems have a characteristic property which will be now discussed. In most cases it is difficult to solve Eqs. (2.30). One way out of the difficulty is sometimes through the transformation from the $\{\vec{p}_i, \vec{q}_i\}$ to another set of variables $\{\vec{P}_i, \vec{Q}_i\}$, $i = 1,\ldots,N$ in which the equations are simpler. Among all possible transformations, particularly relevant are those such that in the new variables the equations of motion are again of the canonical form. These are called canonical (or contact) transformations. If a system is integrable, there exists a transformation to a special set of new variables (referred to as action-angle variables, and denoted as $\vec{I}_i, \vec{\Theta}_i$, instead of $\vec{P}_i, \vec{Q}_i$, $i = 1,\ldots,N$, respectively) such that the new Hamiltonian H' is a function of only half of the new variables (namely the action variables $\{\vec{I}_i\}$) and does not depend on the remaining 3N variables (the "angles" $\{\vec{\Theta}_i\}$). Since the transformation is canonical, the new equations of motion read

$$\dot{\vec{I}}_k = -\nabla_{\vec{\Theta}_k} H' \equiv 0 \tag{3.1}$$

$$\dot{\vec{\Theta}}_k = \nabla_{\vec{I}_k} H' \doteq \vec{\omega}_k(\{\vec{I}_i\}) \quad , \quad k = 1,\ldots,N \quad . \tag{3.2}$$

The first set of equations, (3.1), is immediately integrated and gives

$$\vec{I}_k(t) = \vec{I}_k(0) = \text{const.} \quad , \quad k = 1,\ldots,N \quad , \tag{3.3}$$

namely the action variables can be chosen in such a way as to be exactly the constants of the motion. On the other hand (3.2) gives

$$\vec{\Theta}_k(t) = \vec{\omega}_k t + \vec{\Theta}_k(0) \quad , \quad k = 1,\dots,N \quad . \tag{3.4}$$

The solutions of the equations of motion are, in other words, cyclic (angular) variables, completely determined by the set of characteristic frequencies

$$\vec{\omega}_k = \vec{\omega}_k(\{\vec{I}_i(0)\}) = \text{const.} \quad , \quad k = 1,\dots,N \quad , \tag{3.5}$$

defined by H' through (3.2) and by the initial conditions through (3.3). It follows also that for an integrable system the surface σ is a 3N-dimensional torus.

In order to prove the previous statements, let us review briefly a few properties of the canonical transformation. If

$$\vec{q}_k = \vec{q}_k(\{\vec{Q}_i , \vec{P}_i\}) \quad ,$$

$$\vec{p}_k = \vec{p}_k(\{\vec{Q}_i , \vec{P}_i\}) \quad , \quad k,i = 1,\dots,N \tag{3.6}$$

is a canonical transformation from the set $\{\vec{q}_k , \vec{p}_k\}$ to a set $\{\vec{Q}_i , \vec{P}_i\}$, the equations of motion in the latter will be

$$\dot{\vec{Q}}_i = \nabla_{\vec{P}_i} H' \quad ,$$

$$\dot{\vec{P}}_i = - \nabla_{\vec{Q}_i} H' \quad , \quad i = 1,\dots,N \tag{3.7}$$

where $H' = H'(\{\vec{Q}_i , \vec{P}_i\})$ is the Hamilton function in the new variables. A necessary and sufficient condition for (3.6) to be a canonical transformation is that there exists a generating function $W(\{\vec{q}_k , \vec{P}_k\})$ such that

$$\vec{p}_k = \nabla_{\vec{q}_k} W \quad ,$$

$$\vec{Q}_k = - \nabla_{\vec{P}_k} W \quad , \quad k = 1,\dots,N \quad . \tag{3.8}$$

Indeed, first of all, it follows from the principles of Variational Calculus that

$$\delta \frac{dW}{dt} - \frac{d}{dt}\,\delta W = 0 \quad , \tag{3.9}$$

which, by (3.8), can be cast in the form

$$- \sum_{k=1}^{N} \dot{\vec{p}}_k \cdot \delta\vec{q}_k + \sum_{k=1}^{N} \dot{\vec{q}}_k \cdot \delta\vec{p}_k - \sum_{k=1}^{N} \dot{\vec{Q}}_k \cdot \delta\vec{P}_k +$$

$$+ \sum_{k=1}^{N} \dot{\vec{P}}_k \cdot \delta\vec{Q}_k = 0 \quad . \tag{3.10}$$

From (2.30), on the other hand, one has

$$\delta H = - \sum_{k=1}^{N} \dot{\vec{p}}_k \cdot \delta\vec{q}_k + \sum_{k=1}^{N} \dot{\vec{q}}_k \cdot \delta\vec{p}_k \quad , \tag{3.11}$$

and analogously from (3.7),

$$\delta H' = \sum_{k=1}^{N} \dot{\vec{Q}}_k \cdot \delta\vec{P}_k - \sum_{k=1}^{N} \dot{\vec{P}}_k \cdot \delta\vec{Q}_k \quad . \tag{3.12}$$

Combining (3.11), (3.12) and (3.10) one gets

$$\delta H = \delta H' \tag{3.13}$$

from which there follows immediately $H = H'$, since they are the same quantity (total energy of the system) just expressed in different sets of coordinates. Now, from the principle of stationary action, if the transformation (3.6) is canonical, the two equations

$$\delta \int_{t_1}^{t_2} \left(\sum_{k=1}^{N} \vec{p}_k \cdot \dot{\vec{q}}_k - H \right) dt = 0 \tag{3.14}$$

$$\delta \int_{t_1}^{t_2} \left(\sum_{k=1}^{N} \vec{P}_k \cdot \dot{\vec{Q}}_k - H' \right) dt = 0 \tag{3.15}$$

should hold at the same time. This happens only if

$$\sum_{k=1}^{N} \vec{p}_k \cdot \dot{\vec{q}}_k - H = \sum_{k=1}^{N} \vec{P}_k \cdot \dot{\vec{Q}}_k - H' + \frac{d}{dt} W\left(\{\vec{q}_i , \vec{P}_i\} \right) \tag{3.16}$$

from which (3.8) follow.

The task is finally to find a generating function, which we denote $S = S(\{\vec{q}_i , \vec{I}_i\})$, such that the transformation (3.8),

$$\vec{p}_i = \nabla_{\vec{q}_i} S \quad , \quad \vec{\Theta}_i = \nabla_{\vec{I}_i} S \quad , \quad i = 1,\ldots,N \tag{3.17}$$

changes the Hamiltonian $H(\{\vec{q}_i , \vec{p}_i\})$ into an $H'(\{\vec{I}_i\})$ which is independent of the $\vec{\Theta}_i$'s. This implies that S must satisfy the following partial differential equation

$$H\left(\{\nabla_{\vec{q}_k} S , \vec{q}_k\} \right) = H'\left(\{\vec{I}_k\} \right) = \mathscr{E} \tag{3.18}$$

where $\mathscr{E} = H'\left(\{\vec{I}_k(0)\} \right)$ is the total energy of the system. Equation (3.18) is the Hamilton-Jacobi equation.

It is often difficult to solve the Hamilton-Jacobi equation, but once one has found S the solution of the transformed equations of motion is trivial and is given by (3.3), (3.4). The physical meaning of S is easily seen by evaluating

$$\frac{dS}{dt} = \sum_{k=1}^{N} \nabla_{\vec{q}_k} S \cdot \dot{\vec{q}}_k + \sum_{k=1}^{N} \nabla_{\vec{I}_k} S \cdot \dot{\vec{I}}_k = \sum_{k=1}^{N} \vec{p}_k \cdot \dot{\vec{q}}_k \tag{3.19}$$

due to (3.1) and (3.17). From (3.19)

$$S = \int^{t} \sum_{k=1}^{N} \vec{p}_k \cdot \dot{\vec{q}}_k \, dt \quad . \tag{3.20}$$

Notice that our previous discussion holds only for the most frequent case when the Hamiltonian H does not contain time explicitly.

If it does, one should use a generating function $\bar{S} = \bar{S}(\{\vec{q}_i, \vec{I}_i\}; t)$, and repeat the whole procedure. This replaces Eq. (3.18) with

$$H\left(\{\nabla_{\vec{q}_k} \bar{S}, \vec{q}_k\}, t\right) + \frac{\partial \bar{S}}{\partial t} = 0 \tag{3.21}$$

(where the solution with $H' = 0$ was selected) and (3.20) with

$$\bar{S} = \int^t \left(\sum_{k=1}^{N} \vec{p}_k \cdot d\vec{q}_k - H dt \right) = \int^t \mathscr{L} dt \quad . \tag{3.22}$$

The form $\left(\sum_{k=1}^{N} \vec{p}_k \cdot d\vec{q}_k - H dt \right)$ is the so-called integral invariant of Poincaré - Cartan (and its integral, the integral invariant of Hilbert). If H does not contain time explicitly, (3.22) leads to

$$\bar{S} = S - \mathscr{E} \cdot t \quad . \tag{3.23}$$

If the system is integrable, namely if there exist 3N regular independent functions,

$$J_\alpha = J_\alpha(\{\vec{q}_i, \vec{p}_i\}) \quad , \quad \alpha = 1, \ldots, 3N \quad , \tag{3.24}$$

which remain constant along every trajectory of the system in phase space Φ (for a conservative system, H coincides, of course, with one of the J_α's), and one denotes by $\{c_\alpha\}$ the numerical values of such constants,

$$c_\alpha = J_\alpha\left(\{\vec{q}_i^{(0)}, \vec{p}_i^{(0)}\}\right) \quad , \quad \alpha = 1, \ldots, 3N \quad , \tag{3.25}$$

the 3N equations

$$J_\alpha\left(\{\vec{q}_i\,,\vec{p}_i\}\right) = c_\alpha \qquad , \qquad \alpha = 1,\ldots,3N \qquad , \tag{3.26}$$

can be solved in the form

$$\vec{p}_i = \vec{\pi}_i\,(\{\vec{q}_k\}\,,\{c_\alpha\}) \qquad , \qquad i = 1,\ldots,N \qquad . \tag{3.27}$$

Replacing the latter in the equations of motion

$$\dot{\vec{q}}_i = \nabla_{\vec{p}_i}\, H\left(\{\vec{q}_i\,,\vec{\pi}_i(\{\vec{q}_k\}\,,\{c_\alpha\})\}\right) \qquad , \qquad i = 1,\ldots,N \tag{3.28}$$

these should be solvable by simple integration. Since there are 3N regular independent functions which are constants on the manifold of solutions of the equations of motion, namely (see (3.3)) the $(\vec{I}_i)_\xi$; $i = 1,\ldots,N$; $\xi = 1,2,3$, it is always possible to identify these with the J_α's $\alpha = 1,\ldots,3N$ in a one-to-one way, and the equations (3.26) with the transformation generated by the generating function (3.20) (or (3.22)).

Let $\{u_r\,;\, r = 1,\ldots,\nu\}$ be a set of ν functions of the 2n $(n = 3N)$ variables $\{(\vec{q}_i)_\xi\,,\ (\vec{p}_i)_\xi;\ \ i = 1,\ldots,N,\ \ \xi = 1,2,3\} \equiv \{q_\alpha,\ p_\alpha;\ \ \alpha = 1,\ldots,n\}$. If it is possible to express all the Poisson brackets $\{u_r\,,u_s\}$ as functions of the $\{u_r\}$, the latter set is said to form a function group (the concept was introduced by Sophus Lie). In particular if

$$\{u_r\,,u_s\} = 0 \qquad , \qquad \forall\ r,s = 1,\ldots,\nu \tag{3.29}$$

the functions $\{u_r,\ \ r = 1,\ldots,\nu\}$ are said to be in involution. Now assume that the u_r's are in involution. If the functional equations

$$v_r = 0 \qquad , \qquad r = 1,\ldots,\nu \tag{3.30-a}$$

are consequences of the equations

$$u_r = 0 \quad , \quad r = 1,\ldots,\nu \tag{3.30-b}$$

then also the functions $\{v_r;\ r = 1,\ldots,\nu\}$ are in involution:

$$\{v_r , v_s\} = 0 \quad , \quad \forall\ r,s = 1,\ldots,\nu \ . \tag{3.31}$$

This can be proved by noticing that if f, g are two arbitrary functions in the set $\{v_r\}$, by the hypothesis they must depend on $\{q_\alpha , p_\alpha\}$ only through the $\{u_r\}$. Therefore,

$$\{f , g\} = \frac{1}{2}\sum_{r,s=1}^{\nu}\left(\frac{\partial f}{\partial u_r}\frac{\partial g}{\partial u_s} - \frac{\partial f}{\partial u_s}\frac{\partial g}{\partial u_r}\right)\{u_r , u_s\} = 0 \tag{3.32}$$

where (3.29) was used.

Now one can show that the J_α's are in involution. Indeed, by Jacobi's identity,

$$\{H , \{J_\alpha , J_\beta\}\} = \{\{J_\beta , H\} , J_\alpha\} - \{\{J_\alpha , H\} , J_\beta\} \ ,$$
$$\alpha,\beta = 1,\ldots,3N \quad , \tag{3.33}$$

thus if J_α , J_β are constants of the motion, $\{J_\alpha , J_\beta\}$ is a constant of the motion as well. On the other hand, using the canonical equations of motion, and (3.27),

$$-\frac{\partial H}{\partial q_\alpha} = \dot{p} = \frac{d}{dt}\pi_\alpha = \frac{\partial \pi_\alpha}{\partial t} + \sum_{\beta=1}^{n}\frac{\partial \pi_\alpha}{\partial q_\beta}\dot{q}^\alpha_\beta$$

$$= \frac{\partial \pi_\alpha}{\partial t} + \sum_{\beta=1}^{n}\frac{\partial \pi_\alpha}{\partial q_\beta}\frac{\partial H}{\partial p_\beta} \ . \tag{3.34}$$

However, resorting to the Poisson bracket

$$\{(p_\alpha - \pi_\alpha)\, ,\, (p_\beta - \pi_\beta)\} = \frac{\partial \pi_\beta}{\partial q_\alpha} - \frac{\partial \pi_\alpha}{\partial q_\beta} \quad , \tag{3.35}$$

Eq. (3.34) can be written

$$\frac{\partial \pi_\alpha}{\partial t} = -\frac{\partial H}{\partial q_\alpha} - \sum_{\beta=1}^{n} \frac{\partial H}{\partial p_\beta}\frac{\partial \pi_\beta}{\partial q_\alpha} + \sum_{\beta=1}^{n} \frac{\partial H}{\partial p_\beta}\{(p_\alpha - \pi_\alpha)\, ,\, (p_\beta - \pi_\beta)\}$$

$$\doteq -\frac{\partial \tilde{H}}{\partial q_\alpha} + \sum_{\beta=1}^{n} \frac{\partial H}{\partial p_\beta}\{(p_\alpha - \pi_\alpha)\, ,\, (p_\beta - \pi_\beta)\} \quad , \tag{3.36}$$

where $\tilde{H}$ is the Hamilton's function H, expressed in terms of the arguments $(\{q_\alpha\, ,\, c_\alpha\}\, ;\, t)$. Now, the differential form of Eq. (3.28) is

$$\sum_{\alpha=1}^{n} \pi_\alpha\, dq_\alpha - \tilde{H}\, dt \tag{3.37}$$

which — for integrability — should be a perfect differential of some function $G(\{q_\alpha\, ,\, c_\alpha\}\, ;\, t)$. Equations (3.35), (3.36) show that this is true if and only if the $\{(p_\alpha - \pi_\alpha);\ \alpha = 1,\ldots,n\}$ are in involution,

$$\{(p_\alpha - \pi_\alpha)\quad ,\quad (p_\beta - \pi_\beta)\} = 0 \quad , \quad \forall\ \alpha,\beta = 1,\ldots,n \quad . \tag{3.38}$$

Indeed in such case (3.35), (3.36) become

$$\frac{\partial \pi_\beta}{\partial q_\alpha} = \frac{\partial \pi_\alpha}{\partial q_\beta} \quad , \quad \frac{\partial \pi_\alpha}{\partial t} = -\frac{\partial \tilde{H}}{\partial q_\alpha} \quad , \quad \forall\ \alpha,\beta = 1,\ldots,n \quad . \tag{3.39}$$

[Notice that if G is constructed, then the solution of the equations of motion (3.28) is simply

$$\frac{\partial G}{\partial c_\alpha} = b_\alpha = \text{const.} \tag{3.40}$$

for some arbitrary set of constants $\{b_\alpha;\ \alpha = 1,\dots,n\}$.] Finally, since by definition (see Eqs. (3.26), (3.27)), $p_\alpha = \pi_\alpha$ implies $J_\alpha = c_\alpha$, $\alpha = 1,\dots,n$, upon setting $u_\alpha = p_\alpha - \pi_\alpha$, $v_\alpha = J_\alpha - c_\alpha$, $\nu = n$, the property described by (3.30-a,b) and (3.31) leads to

$$\{J_\alpha , J_\beta\} \equiv 0 \quad , \quad \forall\ \alpha,\beta = 1,\dots,n \ . \tag{3.41}$$

In order to study the topological properties of the manifold σ, let us now consider the 2n-component vectors $\{\vec{v}_\alpha,\ \alpha = 1,\dots,n\}$ in Φ,

$$\vec{v}_\alpha \equiv \left(\frac{\partial J_\alpha}{\partial p_1},\dots,\frac{\partial J_\alpha}{\partial p_n}, -\frac{\partial J_\alpha}{\partial q_1},\dots,-\frac{\partial J_\alpha}{\partial q_n}\right) \quad , \quad \alpha = 1,\dots,n \ . \tag{3.42}$$

Since by (3.41)

$$\sum_{\gamma=1}^{n}\left(\frac{\partial J_\alpha}{\partial q_\gamma}\frac{\partial J_\beta}{\partial p_\gamma} - \frac{\partial J_\alpha}{\partial p_\gamma}\frac{\partial J_\beta}{\partial q_\gamma}\right) = 0 \quad , \quad \forall\ \alpha,\beta = 1,\dots,n \quad , \tag{3.43}$$

each $\vec{v}_\alpha$ is perpendicular to the vectors in the set $\{\vec{n}_\beta;\ \beta = 1,\dots,n\}$, defined as

$$\vec{n}_\beta \equiv \left(\frac{\partial J_\beta}{\partial q_1},\dots,\frac{\partial J_\beta}{\partial q_n},\frac{\partial J_\beta}{\partial p_1},\dots,\frac{\partial J_\beta}{\partial p_n}\right) \quad , \quad \beta = 1,\dots,n \ . \tag{3.44}$$

Indeed

$$\vec{v}_\alpha \cdot \vec{n}_\beta \equiv \{J_\alpha , J_\beta\} = 0 \quad , \quad \forall\ \alpha,\beta = 1,\ldots,n \quad . \tag{3.45}$$

On the other hand, σ is defined by the set of equations (3.26) in Φ, and $\{\vec{n}_\alpha,\ \alpha = 1,\ldots,n\}$ are nothing but a congruence of vectors perpendicular to σ. Equation (3.45) therefore states that on each hyper-surface σ, defined by fixing the set $\{c_\alpha\}$, the vectors $\vec{v}_\alpha$ constitute a set of n independent regular (because the J_α's are independent and analytic) tangent vector fields to σ itself.

Let us confine our attention to dynamical systems for which the accessible region of phase space is finite. Then σ should be compact and parallelizable. Let us recall that a manifold M is parallelizable if its tangent bundle (i.e. the union of tangent spaces to the manifold itself in all its points, whose dimension is twice the dimension of M) is a direct product. The only n-spheres that are parallelizable are S^1, S^3 and S^7. Otherwise the only compact manifold which admits a tangent vector field with no singular points is the n-torus

$$T^n = \underbrace{T^1 \times T^1 \times \ldots \times T^1}_{n \text{ times}} \tag{3.46}$$

where $T^1 \sim S^1 \sim \{(x_1 , x_2) \in \mathbb{R}^2;\ x_1^2 + x_2^2 = 1\}$ is a circle. This is shown intuitively in Fig. 1 for $n = 2$ (in two dimensions, all connected compact orientable manifolds are spheres with g handles: among these only the torus T^2 $(g = 1)$ admits a tangent vector field with no singularity). $\vec{v}$ is not defined at the poles.

The tori $T^n \sim \sigma$ are called "invariant tori" because any orbit originating on them remains on them indefinitely. In the case of action-angle variables, one can think of the variables $\{\Theta_\alpha;\ \alpha = 1,\ldots,n\}$ as angles, on the torus σ; looking for a canonical transformation $\{q_\alpha , p_\alpha\} \to \{I_\alpha , \Theta_\alpha\}$ such that

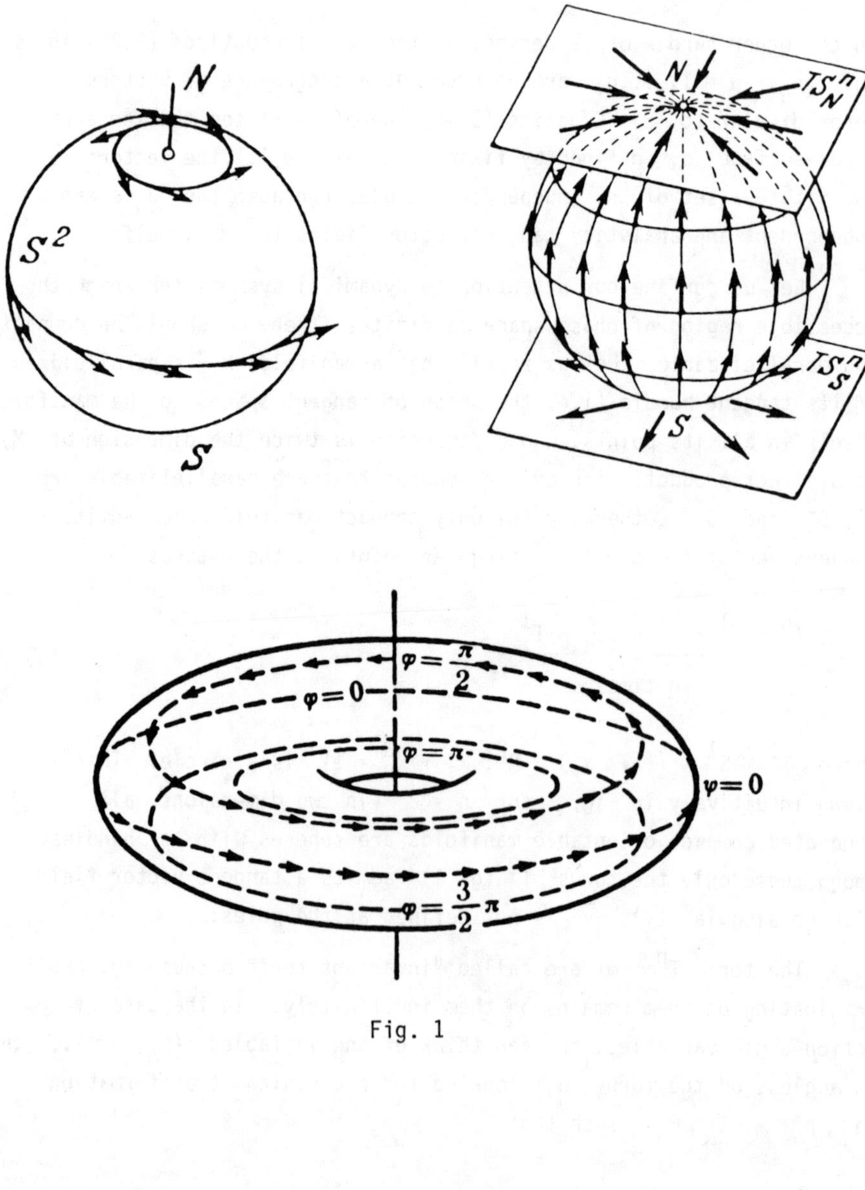
N
S^2
S
N
TS_N^n
TS_S^n
S
$\varphi = \frac{\pi}{2}$
$\varphi = 0$
$\varphi = \pi$
$\psi = 0$
$\varphi = \frac{3}{2}\pi$

Fig. 1

$$I_\alpha = I_\alpha(\{c_\beta\}) \quad , \quad \oint_{\gamma_\alpha} d\Theta_\alpha = 2\pi \quad , \tag{3.47}$$

where $\{\gamma_\alpha, \ \alpha = 1,\ldots,n\}$ are 1-dimensional base cycles on T^n. For $n=1$ Eqs. (3.17), (3.18) give

$$p = \frac{\partial S(q,I)}{\partial q} \quad , \quad \Theta = \frac{\partial S(q,I)}{\partial I} \quad , \quad H\left(\frac{\partial S(q,I)}{\partial q}, q\right) = H'(I) \quad . \tag{3.48}$$

Assuming that the function $H'(I)$ is known and invertible $\gamma = \gamma(I)$ is a closed compact curve in the phase-plane $(q\,, p)$, completely defined by the value of I. From (3.48) one has, at fixed I,

$$dS\Big|_{I = \text{const.}} = p\, dq \quad . \tag{3.49}$$

Thus, upon integration along γ in the neighbourhood of a point q_0,

$$S(q\,, I) = \int_{q_0}^{q} p\, dq \quad . \tag{3.50}$$

Notice that the first requirement in (3.47) is automatically satisfied, $I = I(H') = I(\mathscr{E})$; whereas the second one — being a global requirement — implies more than (3.50). If in the integral (3.50) one goes once around the closed loop γ, returning to q, the integral itself is incremented of the quantity

$$\Delta S(I) = \oint_\gamma p\, dq \doteq A_\gamma \tag{3.51}$$

where A_γ is the area of the portion of $(q\,, p)$-plane which has γ as boundary. Thus S is indeed a multi-valued function on γ, defined up to integral multiples of A_γ. This does not influence the value of $\frac{\partial S}{\partial q}$, but it is reflected on the multi-valuedness of Θ which is then defined up to an integer multiple of $\frac{d}{dt}\Delta S$. In order to satisfy the second condition in (3.47), namely $\oint_{\gamma} d\Theta = 2\pi$, it is then necessary that

$$\frac{d}{dI}\Delta S = 2\pi \quad , \tag{3.52-a}$$

namely

$$I = \frac{\Delta S}{2\pi} = \frac{A_{\gamma}}{2\pi} \quad . \tag{3.52-b}$$

For n degrees of freedom, the whole discussion can be repeated leading to

$$I_{\alpha}(\{c_{\beta}\}) = \frac{1}{2\pi} \oint_{\gamma_{\alpha}} \sum_{\beta=1}^{n} p_{\beta}\, dq_{\beta} \quad , \quad \alpha = 1,\ldots,n \quad . \tag{3.53}$$

Now, one should notice two relevant facts. First of all, due to Stoke's theorem the integral in (3.53) is independent of the choice of the set of representative curves $\{\gamma_{\alpha},\ \alpha = 1,\ldots,n\}$ assumed as base cycles (any other set of curves $\{\gamma'_{\alpha},\ \alpha = 1,\ldots,n\}$ where γ'_{α} belongs to the same homotopy class as γ_{α}, $\forall\alpha$, would give the same I_{α}'s). Second, in a simply connected neighbourhood of a point $\{q_{\alpha}^{(0)}\} \in \sigma$

$$S(\{q_{\alpha}, I_{\alpha}\}) = \int_{\{q_{\alpha}^{(0)}\}}^{\{q_{\alpha}\}} \sum_{\beta=1}^{n} p_{\beta}(\{q_{\gamma}, I_{\gamma}\})\, dq_{\beta} \quad . \tag{3.54}$$

This defines, according to (3.17), a canonical transformation to the action-angle variables in which the coordinates Θ_{α}, $\alpha = 1,\ldots,n$ are multi-valued, with periods

$$\Delta_\beta \Theta_\alpha = \Delta_\beta \frac{\partial S}{\partial I_\alpha} = \frac{\partial}{\partial I_\alpha} \Delta_\beta S =$$

$$= \frac{\partial}{\partial I_\alpha} \oint_{\gamma_\beta} dS = \frac{\partial}{\partial I_\alpha} 2\pi I_\beta = 2\pi \delta_{\alpha\beta} \quad , \quad \alpha,\beta = 1,\ldots,n \quad . \tag{3.55}$$

Equation (3.55) is consistent with the geometrical interpretation of $\{\Theta_\alpha\}$: the increment of coordinate Θ_α along the base cycle γ_β equals 2π if $\beta = \alpha$, zero if $\beta \neq \alpha$; the Θ_α's are just the angular coordinates on T^n.

In conclusion notice also that all the operations performed include only "algebraic" operations (more precisely inversion of functions) and a single quadrature (the integral of a known function) as claimed. Returning to the original variables $\{\vec{q}_i , \vec{p}_i ; \ i = 1,\ldots,N\}$ one can now check that they are periodic functions of $\{\vec{\Theta}_i\}$ with periods $\{2\pi\}$:

$$\vec{q}_j = \sum_{\{\vec{m}_l\}} \vec{Q}^{(j)}_{\{\vec{m}_l\}} (\{\vec{I}_k\}) \exp\left[i \sum_{l=1}^{N} \vec{m}_l \cdot \vec{\Theta}_l\right] \quad ,$$

$$\vec{p}_j = \sum_{\{\vec{m}_l\}} \vec{P}^{(j)}_{\{\vec{m}_l\}} (\{\vec{I}_k\}) \exp\left[i \sum_{l=1}^{N} \vec{m}_l \cdot \vec{\Theta}_l\right] \quad , \quad j = 1,\ldots,N \tag{3.56}$$

where $\vec{Q}^{(j)}_{\{\vec{m}_l\}}$, $\vec{P}^{(j)}_{\{\vec{m}_l\}}$ denote the Fourier amplitudes, and $\{\vec{m}_l ; \ l = 1,\ldots,N\}$ is a set of N three-dimensional vectors with integer components. Using the solutions (3.4), Eqs. (3.56) can be written

$$\vec{q}_j = \vec{q}_j(t) = \sum_{\{\vec{m}_l\}} \vec{Q}^{(j)}_{\{\vec{m}_l\}} (\{\vec{I}_k\}) \exp\left[i \sum_{l=1}^{N} \vec{m}_l \cdot \left(\vec{\omega}_l(\{\vec{I}_k\})t + \vec{\Theta}_l(0)\right) \right] \quad ,$$

$$\vec{p}_j = \vec{p}_j(t) = \sum_{\{\vec{m}_l\}} \vec{p}^{(j)}_{\{\vec{m}_l\}}(\{\vec{I}_k\})\exp\left[i \sum_{l=1}^{N} \vec{m}_l\cdot\left(\vec{\omega}_l(\{\vec{I}_k\})t + \vec{\Theta}_l(0)\right)\right] \quad ,$$

$$j = 1,\ldots,N \quad . \tag{3.57}$$

Thus a generic orbit on σ is multi-periodic, with 3N periods

$$T_\alpha = \frac{2\pi}{\omega_\alpha} \quad , \qquad \alpha = 1,\ldots,3N \quad , \tag{3.58}$$

where $\{\omega_\alpha\} \equiv \{(\vec{\omega}_l)_\xi;\ l = 1,\ldots,N;\ \xi = 1,2,3\}$, each corresponding to a cycle on σ homotopic to γ_α. If there exists a $t = \tau$ such that

$$\vec{\omega}_l(\{\vec{I}_k\})\tau = 2\pi\,\vec{n}_l \quad , \qquad \forall l = 1,\ldots,N \tag{3.59}$$

where $\vec{n}_l$, $l = 1,\ldots,N$ are vectors with integer components, (i.e. the periods T_α are commensurable; namely there exists a set of integers $\{n'_\alpha\}$ such that $\tau = n_\alpha T_\alpha$, $\forall\alpha = 1,\ldots,3N$), then the orbit does not cover σ, but constitutes just a one-dimensional submanifold of σ (see Fig. 2).

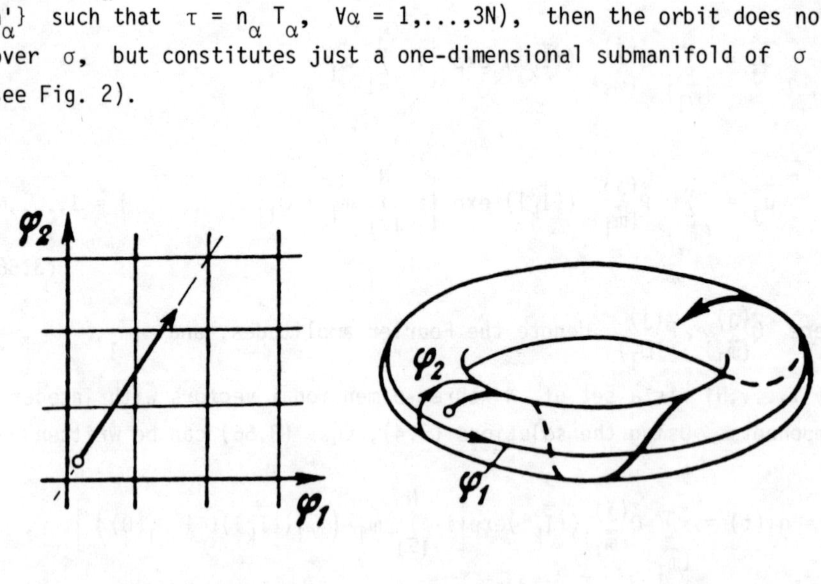

Fig. 2

If the frequencies $\{\omega_\alpha\}$ are incommensurable, the orbit never closes and forms a tight helical path on σ, which eventually (for very large t) covers σ densely. The latter property is sometimes improperly referred to as ergodicity on σ. Indeed the concept of ergodicity (on which we shall return for a thorough discussion in the sequel) refers to that property of a dynamical system of showing characteristic orbits which cover uniformly and densely the energy hypersurface Σ in Φ (namely the surface defined by the equation $H = \mathscr{E}$). Now σ is a submanifold of Σ of dimension $n = dN$ where d is the dimension of the configuration space, whereas Σ has dimension $(2dN - 1)$. Only in the case $N = 1$, $d = 1$ might the two varieties coincide (a single particle in a one-dimensional configuration space, or a system with one degree of freedom), but then $[\dim\Phi = 2;\ \Sigma \sim \sigma \sim T' \sim S'$, a circle$]$ the very concept of ergodicity becomes trivial.

4. Non-Integrable Systems

If the conditions stated at the beginning of Section 3 are not met, namely either the number of constants of motion is $< 3N$ or some (or all) of them are non-analytic (discontinuous, multi-valued, etc.), the action-angle variables cannot be constructed, and the dynamical system is said to be non-integrable.

We will see that then the Hamilton-Jacobi equation has no global solution in terms of differentiable functions and the standard perturbation theory is inapplicable, leading to divergencies which can be controlled with difficulty. The most interesting feature of non-integrable dynamical systems is however that in phase space there appear regions in which the motion exhibits a chaotic character, and the dynamical variables essentially behave as random variables.

Let us consider first a small perturbation of an integrable system. Let $\{I_\alpha, \Theta_\alpha; \alpha = 1,\dots,N\}$ be the action-angle coordinates for the unperturbed system, given by a Hamiltonian $H_0 = H_0(\{I_\alpha\})$ and assume that the perturbed system is given by

$$H = H(\{I_\alpha, \Theta_\alpha\}; \varepsilon) \doteq$$

$$\doteq H_0(\{I_\alpha\}) + \varepsilon H_1(\{I_\alpha, \Theta_\alpha\}) \tag{4.1}$$

where ε is a small parameter $(\varepsilon \ll 1)$ and H_1 is assumed to be an analytic function of its 2n arguments. Naturally $\{I_\alpha, \Theta_\alpha\}$ are good canonical coordinates for the system, but are no longer action-angle variables since the $\{\Theta_\alpha\}$'s appear as arguments of H. Thus the $\{I_\alpha\}$ are not conserved quantities.

For $\varepsilon = 0$ the phase-space Φ $(\dim\phi = 2n)$ is foliated into an n-parameter family of invariant tori $J_\alpha \equiv I_\alpha = c_\alpha$, $\alpha = 1,\ldots,n$. Moreover, by (3.47), (3.56), H as well is a periodic function of $\{\Theta_\alpha\}$, with periods 2π. The question addressed now is: what happens to this foliation for $\varepsilon \neq 0$ and small? If the tori existed also for the perturbed system, one should be able to find a canonical transformation to a new set of variables $(\{I'_\alpha, \Theta'_\alpha\})$, playing the role of action-angle variables.

Introducing once more for simplicity the n-vector notation $(\vec{I} \equiv (I_1,\ldots,I_n), \vec{\Theta} \equiv (\Theta_1,\ldots,\Theta_n))$, the latter should be generated by a function $S = S(\vec{\Theta}, \vec{I}')$ which is a solution of the Hamilton-Jacobi equation (3.18),

$$H(\nabla_{\vec{\Theta}} S(\vec{\Theta}, \vec{I}'), \vec{\Theta}) = H'(\vec{I}') \quad . \tag{4.2}$$

The form (4.1) of H suggests looking for a solution of (4.2) in the form of a formal power series expansion in the perturbative parameter ε. For $\varepsilon = 0$, one should have, of course, $S = S_0 = \vec{\Theta} \cdot \vec{I}'$, which, by Eqs. (3.17), now cast in the form

$$\vec{I} = \nabla_{\vec{\Theta}} S \quad , \quad \vec{\Theta}' = \nabla_{\vec{I}'} S \quad , \tag{4.3}$$

generates the identity $\vec{I}' = \vec{I}$, $\vec{\Theta}' = \vec{\Theta}$.

Let us write therefore (Birkhoff)

$$S = \vec{\Theta}\cdot\vec{I}' + \sum_{k=1}^{\infty} S_k(\vec{\Theta}, \vec{I}')\varepsilon^k \quad . \tag{4.4}$$

Equation (4.2) is now written as

$$H_0\left(\vec{I}' + \sum_{k=1}^{\infty} \varepsilon^k \nabla_{\vec{\Theta}} S_k\right) + \varepsilon H_1\left(\vec{I}' + \sum_{k=1}^{\infty} \varepsilon^k \nabla_{\vec{\Theta}} S_k , \vec{\Theta}\right) = H'(\vec{I}') \quad . \tag{4.5}$$

The idea is to compare on both sides of (4.5) terms corresponding to equal powers of ε to get a set of equations whereby both the $\{S_k\}$ and H' can be obtained. Since all the elements necessary for the discussion to follow are already contained in the terms of the expansion up to the second order, all factors $O(\varepsilon^3)$ will be disregarded. Moreover, denoting by $\vec{\omega}_0 = \nabla_{\vec{I}} H_0 = \vec{\omega}_0(\vec{I})$ the vector of normal frequencies for the unperturbed motion and recalling that — as mentioned before — H_1 is analytic in $\vec{q}, \vec{p}$, whereas the latter are periodic in $\vec{\Theta}$ (Eqs. (3.56)), one can set

$$H_1(\vec{I}, \vec{\Theta}) = \sum_{\vec{m}} H_{1,\vec{m}}(\vec{I}) \exp(i\, \vec{m}\cdot\vec{\Theta}) \quad , \tag{4.6}$$

$$\vec{\Theta} = \vec{\omega}_0 t + \vec{\Theta}_0 \quad , \tag{4.7}$$

where $\vec{m}$ is an n-vector with integer components. The same holds, of course, for $S_k(\vec{\Theta}, \vec{I}')$ (only $k = 1,2$ will be considered) and one writes

$$S_k(\vec{\Theta}, \vec{I}') = \sum_{\substack{\vec{m} \\ \vec{m}\neq\vec{0}}} S_{k,\vec{m}}(\vec{I}') \exp(i\, \vec{m}\cdot\vec{\Theta}) \quad , \tag{4.8}$$

where, since only $\{\nabla_{\vec{\Theta}} S_k\}$'s enter (4.5) the constant term $(\vec{m} = \vec{0})$ of the Fourier series has been set equal to zero.

Equation (4.5) becomes, first

$$H_0(\vec{I}') + \varepsilon\left(\vec{\omega}_0(\vec{I}')\cdot\nabla_{\vec{\Theta}} S_1 + H_1(\vec{I}',\vec{\Theta})\right) +$$

$$+\ \varepsilon^2\left(\vec{\omega}_0(\vec{I}')\cdot\nabla_{\vec{\Theta}} S_2 + \nabla_{\vec{I}} H_1\Big|_{\vec{I}=\vec{I}'}\cdot\nabla_{\vec{\Theta}} S_1 +\right.$$

$$\left.+\ \frac{1}{2}\sum_{\alpha,\beta=1}^{n}\frac{\partial^2 H_0}{\partial I_\alpha \partial I_\beta}\Big|_{\vec{I}=\vec{I}'}\frac{\partial S_1}{\partial\Theta_\alpha}\frac{\partial S_1}{\partial\Theta_\beta}\right) +$$

$$+\ O(\varepsilon^3) = H'(\vec{I}') \quad . \tag{4.9}$$

We assume that the Hessian

$$\mathscr{H} = \det\left\|\frac{\partial^2 H_0}{\partial I_\alpha \partial I_\beta}\right\| \quad , \tag{4.10}$$

does not vanish for $I_\alpha = c_\alpha$, $\alpha = 1,\ldots,n$. Successively, entering (4.6), (4.8) into (4.5) one obtains: (i) upon comparing the term with $\vec{m} = \vec{0}$ in the Fourier expansion,

$$H'(\vec{I}') = H_0(\vec{I}') + \varepsilon H_{1,\vec{0}}(\vec{I}') - \varepsilon^2\left\{ i\sum_{\substack{\vec{m}\\ \vec{m}\neq\vec{0}}} \vec{m}\cdot\nabla_{\vec{I}} H_{1,\vec{m}}\Big|_{\vec{I}=\vec{I}'} S_{1,-\vec{m}}(\vec{I}')\right.$$

$$\left. +\ \frac{1}{2}\sum_{\substack{\vec{m}\\ \vec{m}\neq\vec{0}}}\sum_{\alpha=1}\frac{\partial\omega_{0,\alpha}}{\partial I_\alpha}\Big|_{\vec{I}=\vec{I}'} m_\alpha^2\, S_{1,\vec{m}}(\vec{I}')S_{1,-\vec{m}}(\vec{I}')\right\} + O(\varepsilon^3) \tag{4.11}$$

(ii) for the coefficients of ε, ε^2 on both sides of (4.9) (the term without ε gives obviously an identity), after term-by-term examina- of the Fourier amplitudes of each $\vec{m}\neq\vec{0}$,

$$S_{1,\vec{m}}(\vec{I}') = i \frac{H_{1,\vec{m}}(\vec{I}')}{\vec{m} \cdot \vec{\omega}_0(\vec{I}')} \quad , \tag{4.12}$$

$$S_{2,\vec{m}}(I') = - \frac{i}{\vec{m} \cdot \vec{\omega}_0(\vec{I}')} \sum_{\substack{\vec{l} \\ \vec{l} \neq \vec{0}, \vec{m}}} \left\{ \frac{H_{1,\vec{m}-\vec{l}}(\vec{I}')}{(\vec{m} - \vec{l}) \cdot \vec{\omega}_0(\vec{I}')} \left[\nabla_{\vec{I}} H_{1\vec{l}} \Big|_{\vec{I}=\vec{I}'} \cdot \right. \right.$$

$$\left. \left. \cdot (\vec{m} - \vec{l}) - \frac{1}{2} \frac{H_{1\vec{l}}(\vec{I}')}{\vec{l} \cdot \vec{\omega}_0(\vec{I}')} \sum_{\alpha,\beta=1} l_\alpha (m_\beta - l_\beta) \frac{\partial^2 H_0}{\partial I_\alpha \partial I_\beta} \Big|_{\vec{I}=\vec{I}'} \right] \right\} . \tag{4.13}$$

Insertion of (4.12) into (4.11) gives now the formal expansion of $H'(\vec{I}')$ as a power series in ε, and similarly substitution of (4.12), (4.13) into (4.8) and of the latter into (4.4) provides the analogous expansion of $S = S(\vec{\Theta}, \vec{I}')$ whereby the new action-angles variables $\vec{I}', \vec{\Theta}$ could be generated by (4.3). Thus one would — at first sight — be led to believe that, provided one continued the perturbative scheme up to any arbitrary order in ε, the desired transformation to new action-angle variables could always be found. However this is illusory, as one can already check from the second order expansion. One sees from (4.12), (4.13) that if the frequencies of the unperturbed motion are commensurable (namely they resonate, giving rise — as discussed in (3.59) — to closed orbits on σ), i.e. for some integer-component n-vector $\vec{m}$

$$\vec{\omega}_0 \cdot \vec{m} = 0 \tag{4.14}$$

which case is referred to as "degenerate", the perturbative expansion does not converge (in fact it has no meaning at all).

Moreover, even if this pathological case is excluded, it is well known that an infinite number of vectors $\{\vec{m}\}$ can still be found such that $\vec{\omega}_0 \cdot \vec{m}$ becomes as small as one likes (in particular of order of magnitude less than ε). This has as a result that, for practically all perturbations H_1, the Fourier amplitudes (4.12), (4.13) diverge, as was shown by Poincaré. The corresponding series are, however, indeed semi-convergent; that is, when suitably truncated, these series can predict with great accuracy the behaviour of the system for very long — though not arbitrarily long — periods.

The above problem, which appeared for the first time in celestial mechanics and was known as the "small divisors" problem, leads to a reformulation of the question stated at the beginning of this section in the following way: how should the normal frequencies $\{\omega_\alpha\}$ behave in order that the invariant tori are preserved (modulo small deformations which do not change their topology)? This question is answered by a celebrated theorem, due to Kolmogorov, Arnold and Moser, and named — after their initials — KAM theorem.

Before entering into a discussion of such theorem, we will briefly review some of the connected consequences of non-integrability, and also introduce a different way of representing dynamical systems which will turn out to be particularly useful in what is to follow. It was shown that the single terms of the perturbative series (4.4) do not converge, and, in general, one may gather that the sum of the series itself diverges. Such a behaviour is expected to depend on the fact that one (or more) among the integrals $J_\alpha(\{\vec{q}_i, \vec{p}_i\})$, $\alpha = 1,\ldots,n$ is singular. Indeed this implies that orbits with initial conditions even arbitrarily close to each other in Φ, but on different sides of the singular sub-manifold (and therefore noncontinuable one into the other) may behave quite differently. In other words a characteristic feature of non-integrable systems should be their very sensitive dependence on the choice of initial conditions. Thus the result of integration of the equations of motion can be — in correspondence to certain regions of phase space — quite unpredictable and to some extent chaotic. Notice however that, even though the simplest way to produce a non-integrable system is by adding a non-analytic singular term to a given integrable Hamiltonian (recall that H is one of the J_α's), this is by no means necessary: one can have non-integrable systems with analytic Hamiltonians.

In Eq. (3.22) the Poincaré-Cartan integral-invariant form $\sum_{\alpha=1}^{n} p_\alpha dq_\alpha - H\,dt$ was introduced. On the orbits induced by the solutions of the equations of motion (2.30),

$$\dot{q}_\alpha = \frac{\partial H}{\partial p_\alpha} \quad , \quad \dot{p}_\alpha = - \frac{\partial H}{\partial q_\alpha} \quad , \quad \alpha = 1,\ldots,n \tag{4.15}$$

it should be a perfect integral. There ensues by Stokes' theorem, the following property.

Let $g(t_0, t_1)$ be the transformation (indeed an element of the group of diffeomorphisms of Φ) of the phase space $\Phi \equiv \{q_\alpha, p_\alpha; \alpha = 1,\ldots,n\}$ onto itself induced by (4.15) [namely $g(t_0, t_1) \circ (\{q_\alpha^{(0)}, p_\alpha^{(0)}\}) = (\{q_\alpha(t_1), p_\alpha(t_1)\})$; where $q_\alpha^{(0)} = q_\alpha(t_0), p_\alpha^{(0)} = p_\alpha(t_0)$, $\alpha = 1,\ldots,n$ are the initial conditions and $q_\alpha(t_1), p_\alpha(t_1)$ the corresponding solutions of (4.15) at $t = t_1$]. Let $\gamma \in \Phi \times \mathbb{R}$ be any closed curve in the 2n+1-dimensional direct product space $\Phi \times \mathbb{R}$ ($\mathbb{R}$ denoting the time coordinate) (indeed we shall restrict our attention to curves $\gamma \in \Sigma$, lying on the energy surface Σ at constant time $t = t_0$ (see Fig. 3)). Then $g(t_0, t_1) \circ \gamma$ is itself a closed curve in the same space (lying on the hypersurface of constant energy at $t = t_1$),

$$\oint_\gamma \sum_{\alpha=1}^{n} p_\alpha\, dq_\alpha = \oint_{g(t_0,t_1)\circ\gamma} \sum_{\alpha=1}^{n} p_\alpha\, dq_\alpha \quad , \tag{4.16}$$

and, moreover, any orbit of a point $(\{q_\alpha^{(0)}, p_\alpha^{(0)}\}) \in \gamma$ lies of the flux tube $g(t_0, t) \circ \gamma$, $t_0 \leq t \leq t_1$. There follows also a method of representation of the orbits of a dynamical system (especially useful for non-integrable systems, but, of course, suitable to any system), due to Poincaré, which will be discussed here in the case $n = 2$ for the sake

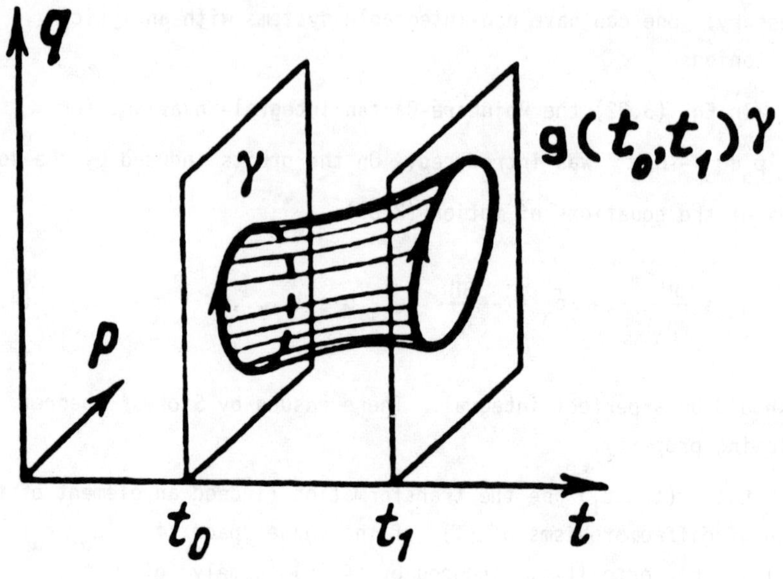

Fig. 3

of graphical description (the extension to any number of degrees of freedom is quite obvious and straightforward).

In this case $\dim \Phi = 4$, whereas $\dim \Sigma = 3$ (notice that Σ is closed, non-Euclidean, possibly multiply-connected). Let's introduce the obvious notation $q_1 \doteqdot x$, $q_2 \doteqdot y$; $p_1 \doteqdot p_x$; $p_2 \doteqdot p_y$, and denote by $\mathscr{S}_x$ the manifold intersection of Σ with the plane $y = 0$. The coordinates on $\mathscr{S}_x$ are (x, p_x). The state of the system is wholly determined (at most up to a sign for p_y if H, as expected, is quadratic in p_y) by a point on $\mathscr{S}_x$. If initial conditions are chosen such that the system "lies" on $\mathscr{S}_x$ at $t = t_0$, one can expect — due to the compactness of Σ — that the representative point will repeatedly traverse $\mathscr{S}_x$ at a sequence of times $t_1, t_2, \ldots, t_k, \ldots$ (namely y will go through the value 0, half of the times with positive p_y, the other half with $p_y < 0$). In other words an orbit initiating in $P_0 \doteqdot (x_0, p_{x_0}) \in \mathscr{S}_x$ at $t = t_0$, will intersect $\mathscr{S}_x$ again and again at

$P_1, P_2 \dots, P_k, \dots,$ where $P_k = g(t_0, t_k) \circ P_0 \doteq (x_k, p_{x,k}) \in \mathscr{S}_x$.

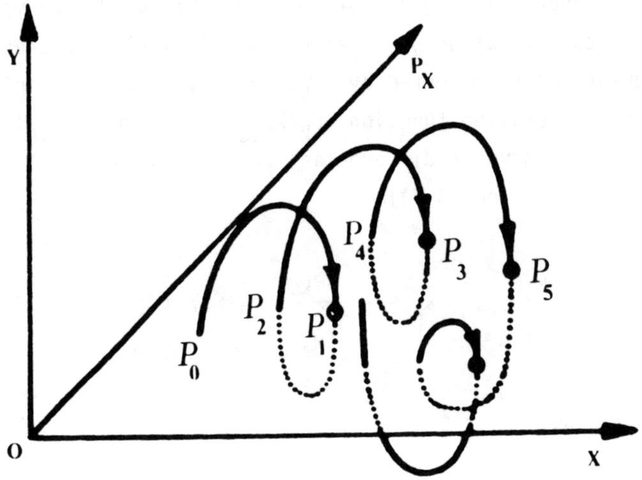

Fig. 4

Of course at any step the whole $\mathscr{S}_x$ is mapped onto itself. Denote by $\mathscr{T}_k$ the map $g(t_{k-1}, t_k)$,

$$\mathscr{T}_k \;:\; P_{k-1} \to P_k \quad ,$$

$$\begin{aligned} P_k &= g(t_{k-1}, t_k) \circ P_{k-1} = \\ &= g(t_{k-1}, t_k)\, g(t_{k-2}, t_{k-1}) \dots g(t_1, t_2)\, g(t_0, t_1) \circ P_0 = \\ &= g(t_0, t_k) \circ P_0 \quad , \end{aligned} \tag{4.17}$$

where the group property of g was used.

Since the equations of motion (4.15) are canonical, the coordinates $\{q_\alpha(t_k), p_\alpha(t_k),\ \alpha = 1,\ldots,n,\ \text{all } k\}$ remain of course canonical and the state characterized by them for a given $k \geq 1$, can be thought of as obtained from the state $\{q_\alpha(t_{k-1}), p_\alpha(t_{k-1})\}$ — considered as initial state — by a canonical transformation. In other words the motion can be thought of as the iteration of a set of (possibly time dependent) canonical transformations represented by $\mathcal{T}_k$, $k = 1,2,\ldots$ and performed by some canonical generating function $S_k(x_{k-1}, p_{x,k};\ t_{k-1}, t_k)$, $k = 1,2,\ldots$ which gives a differential realization of $\mathcal{T}_k$ itself by the usual equations (see (3.8))

$$p_{x,k-1} = \frac{\partial S_k}{\partial x_{k-1}} \quad , \quad x_k = -\frac{\partial S_k}{\partial p_{x,k}} \quad , \quad \forall\ k \geq 1 \quad . \tag{4.18}$$

From (4.18) one has

$$\frac{\partial^2 S_k}{\partial x_{k-1}\partial p_{x,k}} = -\frac{\partial x_k}{\partial x_{k-1}} = \frac{\partial p_{x,k-1}}{\partial p_{x,k}} \quad , \tag{4.19}$$

whereby $\left| j\left((x_{k-1}, p_{x,k-1}) \rightarrow (x_k, p_{x,k})\right)\right| = 1$, with j denoting the Jacobian of the indicated coordinate transformation of $\mathcal{S}_x$. This implies the important property that $\mathcal{T}_k$ preserves the Lebesgue measure over $\mathcal{S}_x$ (the property holds for any number of degrees of freedom: the group of diffeomorphisms generated by the g's preserves the sum of oriented area elements over the n coordinate planes (q_α, p_α), $\alpha = 1,\ldots,n$, of any flux tube elementary section as expected from (4.16)).

The Poincaré representation discussed above is particularly useful in numerical solutions of the equations of motion, in that it allows — by the examination of a sufficiently large number of iterates of a given point P_0 — to decide with a high degree of confidence whether or not the dynamical system under consideration is integrable or not. Indeed, suppose the system is integrable: we have already seen that its orbits do not cover densely the whole Φ but at most the torus σ. The latter intersects $\mathscr{S}_x$ along a closed regular curve Γ, and all $P_k \in \Gamma$. There are thus two possibilities: either the frequencies $\{\omega_\alpha\}$ are commensurate (in the $n=2$ example considered for discussion, there exist two mutually prime integers p, q such that $\omega_1/\omega_2 = p/q$) or they are not (ω_1/ω_2 is an irrational number). In the first case after a sufficiently high number of iterations (such that the corresponding time is multiple integer of $\tau = pT_1 = qT_2$) some iterate (the N-th one, $N \geq pq$) P_N of P_0 coincides with P_0 itself: P_0 is a fixed point of order N of the mapping $\mathscr{T}$. The image of the system consists of N discrete points, belonging to Γ. In the second case no P_k will ever coincide with P_0, but after a large number of iterations the P_k's will clearly exhibit an interpolation of a regular curve, which is just Γ: Γ is an invariant curve for $\mathscr{T}$:

$$\mathscr{T} \circ \Gamma \equiv \Gamma \quad . \tag{4.20}$$

In order to reconstruct the whole Γ in the incommensurate case, one should know at least a portion $\Delta\Gamma$ of it, and repeat the iteration for a sufficient number of orbits with initial points on $\Delta\Gamma$.

For non-integrable systems, the situation is much more complicated: the tori may not exist and the system orbit explores — for a long enough evolution time — a region which is higher dimensional (it could be the whole Σ, if only energy is conserved; in the example considered, Σ is three-dimensional). Thus the intersections of a single orbit with $\mathscr{S}_x$ do not cover a curve but a two-dimensional region. In this case however even a very large number of iterations does not allow any definite statement about the existence of invariant curves (corresponding to invariant tori): indeed it is possible that a set of points apparently random in $\mathscr{S}_x$, lie in fact on a complicated, and yet definite, curve or vice versa that a set of points suggest the existence of a curve from which, by further iterations (i.e. considering a much longer evolution time), the representative point eventually diverges.

A more accurate analysis of the results can be made only in terms of KAM theorem and its consequences, which will be thoroughly discussed in next section. Before proceeding to it, let us, however, add a few more facts about two-dimensional area preserving maps of the form of $\mathscr{T}$ (Eq. (4.17)). Let us write the equations of motion (4.15) in the form

$$\dot{x}_a = X_a \quad , \qquad a = 1,\ldots,2n \quad , \tag{4.21}$$

where x_a, $a = 1,\ldots,2n$ denotes either q_α or p_α (set, e.g. $\alpha = a - n\,[\![\frac{a-1}{n}]\!]$, define $\xi_a \doteq \frac{a-\alpha}{n} \equiv [\![\frac{a-1}{n}]\!]$ and use the convention that $x_a \equiv q_\alpha$ if $\xi_a = 0$, $x_a \equiv p_\alpha$ if $\xi_\alpha = 1$). The $X_a = X_a(\{x_a\};t)$, $a = 1,\ldots,2n$ $\left(X_a = (-)^{\xi_a} \frac{\partial H}{\partial x_a} \; ; \; H = H(\{x_a\}) \right)$ are functions of t that we assume — according to the general discussion — to be periodic of period T. Suppose that periodic solutions of (4.21) are known,

$$x_a = \phi_a(t) \quad , \qquad a = 1,\ldots,2n \quad , \tag{4.22}$$

where

$$\phi_a(t + T) = \phi_a(t) \quad . \tag{4.23}$$

In order to investigate solutions close to (4.22), set

$$x_a = \phi_a(t) + \xi_a \quad , \qquad a = 1,\ldots,2n \quad , \tag{4.24}$$

where $\{\xi_a\}$ are small, given by the first-order differential equations

$$\frac{d\xi_a}{dt} = \sum_{b=1}^{2n} \xi_b \cdot \frac{\partial X_a}{\partial x_b} \quad , \qquad a = 1,\ldots,2n \quad . \tag{4.25}$$

Equations (4.25) are linear differential equations of the first order with coefficients periodic in the independent variable t. Their solutions, according to the general theory of differential equations will be of the form

$$\xi_a = \sum_{b=1}^{2n} S_{ab}(t)\exp(\alpha_b t) \quad , \qquad a, b = 1,\dots,2n \tag{4.26}$$

where

$$S_{ab}(t+T) = S_{ab}(t) \tag{4.27}$$

are periodic functions of t. The $2n$ constants $\{\alpha_a,\ a = 1,\dots,2n\}$ are called characteristic exponents.

If the characteristic exponents are all purely imaginary, the functions $\{\xi_a\}$ given by (4.26) are still periodic (of course with period different from T, unless the α_a's are all zero). For one of the orbits close to a periodic orbit, set $\xi_a(t_0) \doteq \xi_a^{(0)}$; $\xi_a(t_0 + T) \doteq \eta_a + \xi_a^{(0)}$; $a = 1,\dots,2n$ for some arbitrary time $t = t_0$. The $\{\eta_a\}$ are single valued functions of $\{\xi_a^{(0)}\}$, which are zero when the latter are zero. By Taylor's theorem (up to the first order) one can therefore write

$$\eta_a = \sum_{b=1}^{2n} \frac{\partial \eta_a}{\partial \xi_b^{(0)}}\,\xi_b^{(0)} \quad , \qquad a = 1,\dots,2n \quad . \tag{4.28}$$

If α_c is one of the characteristic exponents of the periodic orbit, one of the adjacent orbits will be defined by equations of the form

$$\xi_a = e^{\alpha_c t} S_{ac} \quad , \qquad a = 1,\dots,2n \quad , \tag{4.29}$$

so that

$$\eta_a + \xi_a^{(0)} = e^{\alpha_c T} e^{\alpha_c t_0} S_{ac}(t_0 + T) =$$

$$= e^{\alpha_c T} e^{\alpha_c t_0} S_{ac}(t_0) = e^{\alpha_c T} \xi_a^{(0)} \quad ,$$

$$a = 1,\ldots,2n \quad , \tag{4.30}$$

and (4.28) becomes

$$\sum_{b=1}^{2n} \frac{\partial \eta_a}{\partial \xi_b^{(0)}} \xi_b^{(0)} = (e^{\alpha T} - 1)\, \xi_a^{(0)} \quad , \quad a = 1,\ldots,2n \quad . \tag{4.31}$$

Denoting then by $\{\lambda_c, \ c = 1,\ldots,2n\}$ the eigenvalues of the Jacobian matrix

$$J = \frac{\partial\,(\eta_1,\ldots,\eta_{2n})}{\partial\,(\xi_1^{(0)},\ldots,\xi_{2n}^{(0)})} \quad , \tag{4.32}$$

the characteristic exponents are given by

$$\alpha_a = \frac{1}{T} \ln(1 + \lambda_a) \quad , \quad a = 1,\ldots,2n \quad . \tag{4.33}$$

The above criterion is global in that it gives information about the possible existence of periodic orbits in a neighbourhood of a steady periodic solution of the equations of motion.

Let us recall here also a few basic facts about local stability analysis, which will be particularly relevant in the study of Poincaré's return maps. Since the latter was defined for a two-dimensional manifold $\mathscr{S}_x$, the notation will be referred to a dynamical system with 2 degrees of freedom.

The equations of motion (4.15), written explicitly, read, in this case,

$$\dot{x} = \frac{\partial H}{\partial p_x} \quad , \quad \dot{y} = \frac{\partial H}{\partial p_y} \quad , \quad \dot{p}_x = -\frac{\partial H}{\partial x} \quad , \quad \dot{p}_y = -\frac{\partial H}{\partial y} \tag{4.34}$$

where, for a conservative system,

$$H = H(x\,,y\,,p_x\,,p_y) = E = \text{const.} \tag{4.35}$$

Solving (4.35) in terms of p_y, we have, say,

$$p_y = \Phi_E(x\,,y\,,p_x) \quad . \tag{4.36}$$

Upon defining

$$\left.\frac{\partial H}{\partial p_x}\right|_{p_y=\Phi_E(x,y,p_x)} \doteqdot \tilde{X}(x\,,y\,,p_x) \quad ,$$

$$\left.\frac{\partial H}{\partial p_y}\right|_{p_y=\Phi_E(x,y,p_x)} \doteqdot \tilde{Y}(x\,,y\,,p_x) \quad ,$$

$$\left.\frac{\partial H}{\partial x}\right|_{p_y=\Phi_E(x,y,p_x)} \doteqdot -\,\tilde{P}(x\,,y\,,p_x) \tag{4.37}$$

and writing for simplicity p instead of p_x, one has then, instead of (4.34)

$$\dot{x} = \tilde{X}(x\,,y\,,p) \quad , \quad \dot{y} = \tilde{Y}(x\,,y\,,p) \quad , \quad \dot{p} = \tilde{P}(x\,,y\,,p) \quad . \tag{4.38}$$

Notice that the last equation in (4.34) has been dropped, in that it reduces now to an identity (which guarantees that $\frac{dH}{dt} = 0$). One selects now initial conditions on $\mathscr{S}_x$,

$$x(0) = x_0 \quad , \quad y(0) = 0 \quad , \quad p(0) = p_0 \quad , \tag{4.39}$$

and determines the return time T_1 by the equation

$$y(T_1) = \sum_{m=1}^{\infty} \frac{1}{m!} \left. \frac{d^m y}{dt^m} \right|_{t=0} T_1^m = 0 \quad . \tag{4.40}$$

With the assignments $x \doteq x_1$, $p \doteq x_2$, $y \doteq x_3$, $\tilde{X} \doteq X_1$, $\tilde{P} \doteq X_2$, $\tilde{Y} \doteq X_3$, one has — due to (4.38) —

$$\left. \frac{d^m x_i}{dt^m} \right|_{t=0} = \sum_{j_1=1}^{3} X_{j_1} \frac{\partial}{\partial x_{j_1}} \sum_{j_2=1}^{3} X_{j_2} \frac{\partial}{\partial x_{j_2}} \cdots$$

$$\cdots \sum_{j_m=1}^{3} X_{j_m} \frac{\partial}{\partial x_{j_m}} X_i \bigg|_{t=0} \doteq$$

$$\doteq \psi_i^{(m)}(x_0 , p_0) \quad , \qquad i = 1,2,3 \quad . \tag{4.41}$$

Thus T_1, solution of the equation

$$\sum_{m=1}^{\infty} \frac{1}{m!} \psi_3^{(m)}(x_0 , p_0) T_1^m = 0 \quad , \tag{4.42}$$

is in fact a function of x_0 , p_0; $T_1 = T_1(x_0 , p_0)$.

Once T_1 has been computed, one writes

$$x_1 \doteq x(T_1) = x_0 + \sum_{m=1}^{\infty} \frac{1}{m!} \psi_1^{(m)}(x_0 , p_0)[T_1(x_0 , p_0)]^m \doteq \zeta_1(x_0 , p_0) \ ,$$

$$p_1 \doteq p(T_1) = p_0 + \sum_{m=1}^{\infty} \frac{1}{m!} \psi_2^{(m)}(x_0 , p_0)[T_1(x_0 , p_0)]^m \doteq \zeta_2(x_0 , p_0) \ . \tag{4.43}$$

Finally denote by $\bar{x}_0, \bar{p}_0$ the pair of values, such that

$$\zeta_1(\bar{x}_0, \bar{p}_0) = \bar{x}_0 \quad , \quad \zeta_2(\bar{x}_0, \bar{p}_0) = \bar{p}_0 \tag{4.44}$$

and set

$$x_i = \bar{x}_0(1 + \varepsilon_i) \quad , \quad p_i = \bar{p}_0(1 + \eta_i) \quad , \quad i = 0,1 \quad . \tag{4.45}$$

Clearly $\bar{x}_0, \bar{p}_0$ is a fixed point in $\mathscr{S}_x$ due to the relations (4.44) and we study now the stability of the solutions of (4.43) themselves in proximity of such a steady state by their linear approximation, namely considering ε_i, η_i small $(\varepsilon_i, \eta_i \ll 1)$ and keeping only linear terms (i.e. neglecting terms of $O(\varepsilon_i^2)$, $O(\eta_i^2)$, $O(\varepsilon_i\eta_i)$). One finds

$$\begin{aligned} \varepsilon_1 &= t_{11}\,\varepsilon_0 + t_{12}\,\eta_0 \quad , \\ \eta_1 &= t_{21}\,\varepsilon_0 + t_{22}\,\eta_0 \end{aligned} \tag{4.46}$$

where

$$t_{j1} = \left.\frac{\partial\zeta_j}{\partial x_0}\right|_{\substack{x_0=\bar{x}_0 \\ p_0=\bar{p}_0}} \quad , \quad t_{j2} = \left.\frac{\partial\zeta_j}{\partial p_0}\right|_{\substack{x_0=\bar{x}_0 \\ p_0=\bar{p}_0}} \quad , \quad j = 1,2 \quad . \tag{4.47}$$

Equations (4.46) realize the mapping $\mathscr{T}_1$ of (4.17) in its linearized form. If the system is conservative the mapping is area-preserving and upon writing (4.46) in the matrix form

$$\begin{pmatrix} \varepsilon_1 \\ \eta_1 \end{pmatrix} = \mathscr{T} \begin{pmatrix} \varepsilon_0 \\ \eta_0 \end{pmatrix} \tag{4.48}$$

where $\mathscr{T}$ is the 2×2 matrix of elements $\{t_{ij};\ i,j = 1,2\}$ given by (4.47), one should find

$$\det \mathscr{T} = 1 \quad . \tag{4.49}$$

By the local change of coordinates, in the neighbourhood of the fixed point $(\bar{x}_0, \bar{p}_0)$,

$$\begin{pmatrix} \tilde{\varepsilon}_i \\ \tilde{\eta}_i \end{pmatrix} = R \begin{pmatrix} \varepsilon_i \\ \eta_i \end{pmatrix} \quad , \qquad i = 0,1 \tag{4.50}$$

where R is the 2×2 matrix which diagonalizes $\mathscr{T}$:

$$R \mathscr{T} R^{-1} = \mathrm{diag}\,(\lambda_1, \lambda_2) \tag{4.51}$$

whence

$$\lambda_{1,2} = \frac{1}{2}\left\{ \mathrm{Tr}\,\mathscr{T} \pm \sqrt{(\mathrm{Tr}\,\mathscr{T})^2 - 4 \det \mathscr{T}} \right\} \quad , \tag{4.52}$$

Eq. (4.48) may be written as

$$\tilde{\varepsilon}_1 = \lambda_1 \tilde{\varepsilon}_0 \quad , \qquad \tilde{\eta}_1 = \lambda_2 \tilde{\eta}_0 \quad . \tag{4.53}$$

The whole procedure can now be repeated for any step of the Poincaré mapping (4.17).

In several cases, $\mathscr{T}_k$ as a matrix turns out to be independent of k (and therefore equal to $\mathscr{T}$), and one has the recursive equation

$$\begin{pmatrix} \varepsilon_k \\ \eta_k \end{pmatrix} = \mathscr{T} \begin{pmatrix} \varepsilon_{k-1} \\ \eta_{k-1} \end{pmatrix} \quad , \qquad \forall\ k \geq 1 \tag{4.54}$$

whose solution is

$$\begin{pmatrix} \varepsilon_k \\ \eta_k \end{pmatrix} = \mathscr{T}^k \begin{pmatrix} \varepsilon_0 \\ \eta_0 \end{pmatrix} \quad , \qquad k \geq 1 \tag{4.55}$$

or, equivalently

$$\tilde{\varepsilon}_k = \lambda_1^k \, \tilde{\varepsilon}_0 \quad , \qquad \tilde{\eta}_k = \lambda_2^k \, \tilde{\eta}_0 \quad . \tag{4.56}$$

Naturally, $\mathscr{T}$ is expected not to be too far from the identity, and denoting by μ_j, $j = 1,2$ the eigenvalues of $\mathscr{M} \doteq (\mathscr{T} - \mathbb{I}_2)$; $\mu_j = \lambda_j - 1$; $|\mu_j| << 1$, $j = 1,2$ Eqs. (4.56) give the asymptotic form $(N >> 1)$

$$\begin{pmatrix} \tilde{x}_N \\ \tilde{p}_N \end{pmatrix} \doteq R \begin{pmatrix} x_N \\ p_N \end{pmatrix} = \begin{pmatrix} \bar{x}_0 + (\tilde{x}_0 - \bar{x}_0)e^{\mu_1 N} \\ \bar{p}_0 + (\tilde{p}_0 - \bar{p}_0)e^{\mu_2 N} \end{pmatrix} \quad . \tag{4.57}$$

The eigenvalues μ_1 , μ_2 in general are complex numbers and several situations may occur, depending on the values of their real and imaginary parts.

The completely stable case corresponds to μ_1 and μ_2 both real and negative. The solutions of the iterated map then decay exponentially towards the steady state represented by $(\bar{x}_0 , \bar{p}_0)$, which is called a nodal point. Analogously the completely unstable case corresponds to μ_1 , μ_2 both real and positive. The solutions move away exponentially from the steady state.

An unstable situation occurs also when μ_1 and μ_2 are both real and one is positive, the other negative. This case is referred to as a saddle point (or hyperbolic point). The coordinate line corresponding to the negative eigenvalue is called a separatrix. All other trajectories are diffeomorphic to hyperbolas. If both μ_1 and μ_2 are pure imaginary one has a case of marginal stability: in order to find stability one should consider nonlinear terms. The solutions neither decay to the steady state nor move away from it: they simply oscillate around it, along a closed trajectory diffeomorphic to a circle, remaining in its neighbourhood. The steady state is called a center (or elliptic point). If both μ_1 and μ_2 are complex (of course mutually conjugate), the solution is stable when their real part is negative, unstable when it is positive: in the first case the solutions spiral towards the steady state (which is called a focus), in the second they spiral away from it. Figure 5 illustrates the different possibilities discussed.

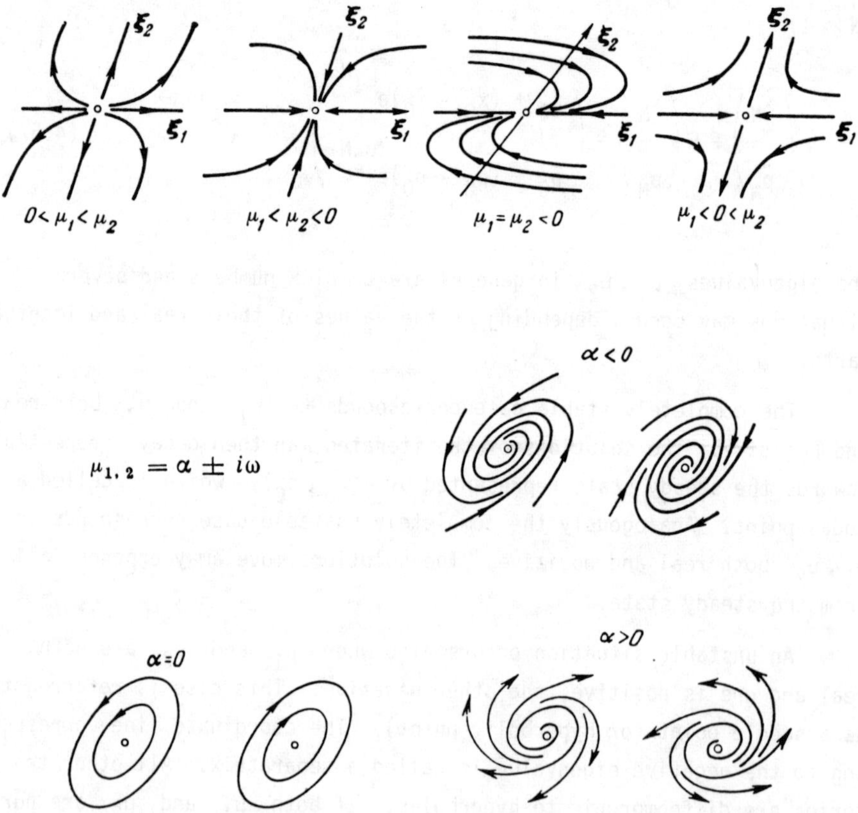

Fig. 5

Notice that the completely stable case can never occur for the area preserving mapping, because $\det \mathcal{T} = 1$ implies $\det \mathcal{M} = -\operatorname{Tr}\mathcal{M}$, which has no two real negative solutions for 2×2 matrices. Moreover, in the saddle point case if $\det \mathcal{T} = 1$, one should have — in the approximation adopted — $\mu_2 = -\mu_1$. One gets then from (4.57)

$$(\tilde{x}_N - \bar{x}_0)(\tilde{p}_N - \bar{p}_0) = \text{const.} = (\tilde{x}_0 - \bar{x}_0)(\tilde{p}_0 - \bar{p}_0) \quad . \tag{4.58}$$

The invariant curves, as mentioned, are hyperbolas. Analogously in the center-point case, it should be $\mu_1 = i\phi = -\mu_2$, and one would have

$$\frac{(\tilde{x}_N - \bar{x}_0)^2}{2(\tilde{x}_0 - \bar{x}_0)^2} + \frac{(\tilde{p}_N - \bar{p}_0)^2}{2(\tilde{p}_0 - \bar{p}_0)^2} = 1 \quad . \tag{4.59}$$

The invariant curves are ellipses. Performing a new rotation around $(\bar{x}_0 , \bar{p}_0)$:

$$\begin{pmatrix} x'_N \\ p'_N \end{pmatrix} = \mathcal{R} \begin{pmatrix} \tilde{x}_N - \bar{x}_0 \\ \tilde{p}_N - \bar{p}_0 \end{pmatrix} \quad , \quad \mathcal{R} = \frac{1}{\sqrt{2}} \begin{pmatrix} 1 & -i \\ i & -1 \end{pmatrix} = \mathcal{R}^{-1} \tag{4.60}$$

they turn into circles:

$$\begin{pmatrix} x'_N \\ p'_N \end{pmatrix} = \begin{pmatrix} \cos N\phi & \sin N\phi \\ -\sin N\phi & \cos N\phi \end{pmatrix} \begin{pmatrix} x'_0 \\ p'_0 \end{pmatrix} \quad , \tag{4.61}$$

$$x'^2_N + p'^2_N = \text{const.} = x'^2_0 + p'^2_0 \quad . \tag{4.62}$$

5. The KAM Theorem

Let us return to Eq. (4.1) that we write in the n-vector notation introduced in (4.2) and in the ensuing discussion:

$$H = H(\vec{I}, \vec{\Theta}, \varepsilon) = H_0(\vec{I}) + \varepsilon H_1(\vec{I}, \vec{\Theta}) \quad . \tag{5.1}$$

For $\varepsilon = 0$, we have an n-parameter family of invariant tori in Φ

$$I_\alpha = c_\alpha \quad , \quad \alpha = 1,\ldots,n \tag{5.2}$$

on which the motion occurs with normal frequencies

$$\vec{\omega}_0 = \nabla_{\vec{I}} H_0 (\vec{c}) \quad . \tag{5.3}$$

The KAM theorem states that for small values of $|\varepsilon|$ those invariant tori of H_0 for which the frequencies $\omega_{0\alpha}$, $\alpha = 1,\ldots,n$ are not only incommensurate but satisfy a condition of the form

$$|\vec{m} \cdot \vec{\omega}_0| \geq a \, \| \vec{m} \|^{-b} \tag{5.4}$$

for any $\vec{m} \neq 0$, where $\vec{m}$ is a vector with integer components such that

$$\| m \| = \sum_{\alpha=1}^{n} |m_\alpha| \geq 1 \tag{5.5}$$

and a, b are positive constants, persist under the perturbation εH_1. In other words the tori are simply slightly deformed in such a way that in the phase space of the perturbed system there are still invariant tori on which the orbits are dense everywhere and are characterized by n frequencies ω_α, $\alpha = 1,\ldots,n$. The condition $\mathscr{H} \neq 0$ for $\vec{I} = \vec{c}$ ($\mathscr{H}$ is defined in (4.10)) is required. The new tori have the form

$$\vec{\Theta} = \vec{\theta} + \vec{u}(\vec{\theta}, \varepsilon) \quad ,$$

$$\vec{I} = \vec{c} + \vec{v}(\vec{\theta}, \varepsilon) \quad , \tag{5.6}$$

where, $\vec{u}(\vec{\theta}, \varepsilon)$ and $\vec{v}(\vec{\theta}, \varepsilon)$ are real analytic functions of $\vec{\theta}$ and ε, with

$$\dot{\vec{\theta}} = \vec{\omega}_0 \quad . \tag{5.7}$$

Moreover they are the majority of the original set of tori, in the sense that the measure of the complement of their union in Φ vanishes with ε.

The proof of KAM theorem is based on two main elements. First, the condition (5.4) defines a range of non-resonant frequencies of the unperturbed system such that its frequencies not only are independent but do not approach — for b large enough — any resonance relation, however small its order. The measure of such a set of frequencies is small with a; and a, of course, vanishes with ε. One successively tries to find in the neighbourhood of any invariant non-resonant torus of the unperturbed system, corresponding to frequencies in the specified range, an invariant torus for the perturbed system which besides being quasi-periodic would be characterized by exactly the same frequencies as the unperturbed system. Thus one sets up a perturbative scheme in which, instead of the frequencies, the initial conditions for the perturbation are varied, in such a way as to keep the frequencies constant and non-resonant. Since the frequencies vary with the action variables (Eqs. (5.2), (5.3)) the condition of non-degeneracy implies that such a variation can be assumed of the order of ε.

The second element is that the search for the invariant torus can be performed by a rapidly convergent method, (which is a generalization of Newton's method of tangents) instead of the customary power series expansion of perturbation theory. Such a method gives, after n iterations, an error of the order ε^{2n}, and the resulting convergence is so fast that one can neutralize the effect of small divisors, and perform — in principle — an infinite number of iterations, thus showing the convergence of the whole procedure.

In order to sketch a proof of the theorem, we restrict our attention to a special class of dynamical systems, called reversible.

Let us write the equations of motion induced by the Hamiltonian (5.1) in the form:

$$\dot{\vec{\Theta}} = \vec{F}(\vec{\Theta}, \vec{I}; \varepsilon) \quad ,$$

$$\dot{\vec{I}} = \vec{G}(\vec{\Theta}, \vec{I}; \varepsilon) \quad , \tag{5.8}$$

where, as stated before, $\vec{F}$ and $\vec{G}$ are real analytic functions, periodic with period 2π in each Θ_α, $\alpha = 1,\ldots,n$, such that

$$\vec{F}(\vec{\Theta}, \vec{I}; 0) = \vec{\mathscr{F}}(\vec{I}) \quad ,$$

$$\vec{G}(\vec{\Theta}, \vec{I}; 0) = \vec{\mathscr{G}}(\vec{I}) \quad . \tag{5.9}$$

Reversible systems are those for which

$$\vec{F}(-\vec{\Theta}, \vec{I}, \varepsilon) = \vec{F}(\vec{\Theta}, \vec{I}, \varepsilon) \quad , \tag{5.10}$$

$$\vec{G}(-\vec{\Theta}, \vec{I}, \varepsilon) = -\vec{G}(\vec{\Theta}, \vec{I}, \varepsilon) \quad . \tag{5.11}$$

Notice that (5.11) implies $\vec{\mathscr{G}}(\vec{I}) \equiv 0$, so that for $\varepsilon = 0$, $\vec{I} = \vec{c}$, are the invariant tori on which the motion is given by

$$\dot{\vec{\Theta}} = \vec{\mathscr{F}}(\vec{c}) \quad . \tag{5.12}$$

In other words $\vec{\mathscr{F}}(\vec{c}) \equiv \omega_0$, and Eq. (4.10),

$$\det \left| \left(\frac{\partial \mathscr{F}_\alpha}{\partial I_\beta} \right) \right|_{\vec{I}=\vec{c}} \neq 0 \tag{5.13}$$

guarantees that the constants c_α, $\alpha = 1,\ldots,n$ can indeed be chosen in such a way that a condition like (5.4) is satisfied. The denomination reversible refers to the fact that the equations of motion (5.8) are invariant under the transformation $\vec{\Theta} \to -\vec{\Theta}$, $t \to -t$.

The proof of the theorem is based on the construction of an iteration scheme which converges so rapidly that it controls the cumulative effect of the small divisors. Consider the group $\mathfrak{G}$ of coordinate transformations

$$\vec{\Theta} = \vec{U}(\vec{\xi}, \vec{\eta})$$

$$\vec{I} = \vec{V}(\vec{\xi}, \vec{\eta}) \qquad (5.14)$$

such that reversibility, real analyticity and periodicity (in half of the new variables) is preserved. Such a group is generated by the requirement that, e.g.,

$$\vec{U}(-\vec{\xi}, \vec{\eta}) = -\vec{U}(\vec{\xi}, \vec{\eta}) \quad ,$$

$$\vec{V}(-\vec{\xi}, \vec{\eta}) = \vec{V}(\vec{\xi}, \vec{\eta}) \quad , \qquad (5.15)$$

and that $\vec{U}(\vec{\xi}, \vec{\eta}) - \vec{\xi}$; $\vec{V}(\vec{\xi}, \vec{\eta})$ are real, analytic in all variables and are periodic with period 2π in ξ_α, $\alpha = 1,\ldots,n$.

We write down the transformation equations in such a way that the new system reads

$$\dot{\vec{\xi}} = \vec{X}(\vec{\xi}, \vec{\eta})$$

$$\dot{\vec{\eta}} = \vec{Y}(\vec{\xi}, \vec{\eta}) \quad . \qquad (5.16)$$

Letting $\vec{U}_{\vec{\xi}}, \vec{U}_{\vec{\eta}}, \vec{V}_{\vec{\xi}}, \vec{V}_{\vec{\eta}}$ denote the matrices of elements $\frac{\partial U_\alpha}{\partial \xi_\beta}$, $\frac{\partial U_\alpha}{\partial \eta_\beta}$, $\frac{\partial V_\alpha}{\partial \xi_\beta}$, $\frac{\partial V_\alpha}{\partial \eta_\beta}$; $\alpha,\beta = 1,\ldots,n$, they read (omitting the argument ε)

$$\vec{U}_{\vec{\xi}} \cdot \vec{X} + \vec{U}_{\vec{\eta}} \cdot \vec{Y} = \vec{F}(\vec{U}, \vec{V}) \quad ,$$

$$\vec{V}_{\vec{\xi}} \cdot \vec{X} + \vec{V}_{\vec{\eta}} \cdot \vec{Y} = \vec{G}(\vec{U}, \vec{V}) \quad . \qquad (5.17)$$

Since $\vec{G}(\vec{\Theta}, \vec{I}, 0) = 0$, one may assume that $\vec{G}(\vec{\Theta}, \vec{I})$ is a small function (of the order of ε): denote it by $\hat{G}$ to remind us of this. Moreover, the unperturbed case suggests that one can set

$$\vec{F}(\vec{\Theta}, \vec{I}) = \vec{\omega}_0 + \vec{I} - \vec{c} + \hat{F} \qquad (5.18)$$

[where $\hat{F}$ is small (of the order of ε)] because this corresponds to assuming $\vec{\mathscr{F}}(\vec{I}) = \vec{\omega}_0 + \vec{I} - \vec{c}$, which satisfies (5.13), and gives $\vec{\mathscr{F}}(\vec{c}) = \vec{\omega}_0$ as desired. It will be shown that such a choice for $\vec{\mathscr{F}}$ is not restrictive of generality. Finally we look for solutions of (5.17) which are linear in $\vec{\eta}$, setting

$$\vec{U} \doteq \vec{\xi} + \vec{U}_0 + U_1 \cdot \vec{\eta} \quad ,$$

$$\vec{V} \doteq \vec{\eta} + \vec{V}_0 + V_1 \cdot \vec{\eta} \quad , \tag{5.19}$$

where $\vec{U}_0, \vec{V}_0$ are n-vectors and $\vec{U}_1, \vec{V}_1$ are $n \times n$ matrices; all are periodic functions of ξ_α, $\alpha = 1,\ldots,n$.

Upon inserting (5.18) and (5.19) in (5.17) one finally obtains, upon comparing on both sides terms independent of $\vec{\eta}$ and linear in $\vec{\eta}$ respectively, the two following systems of non-linear partial differential equations:

$$(\vec{\omega}_0 \cdot \nabla_{\vec{\xi}})\vec{U}_0 - \vec{V}_0 = \hat{F}(\vec{\xi} + \vec{U}_0, \vec{V}_0)$$

$$(\vec{\omega}_0 \cdot \nabla_{\vec{\xi}})\vec{V}_0 = \hat{G}(\vec{\xi} + \vec{U}_0, \vec{V}_0) \tag{5.20}$$

and

$$(\vec{\omega}_0 \cdot \nabla_{\vec{\xi}})U_1 - V_1 = -\vec{U}_{0\vec{\xi}} + \hat{F}_{\vec{\Theta}}(\vec{\xi} + \vec{U}_0, \vec{V}_0)U_1 + {} + \hat{F}_{\vec{I}}(\vec{\xi} + \vec{U}_0, \vec{V}_0)(\mathbb{I}_n + V_1)$$

$$(\vec{\omega}_0 \cdot \nabla_{\vec{\xi}})V_1 = -\vec{V}_{0\vec{\xi}} + \hat{G}_{\vec{\Theta}}(\vec{\xi} + \vec{U}_0, \vec{V}_0)U_1 + \hat{G}_{\vec{I}}(\vec{\xi} + \vec{U}_0, \vec{V}_0)(\mathbb{I}_n + V_1) \tag{5.21}$$

with

$$\vec{X} = \vec{\omega}_0 + \vec{\eta} + O(|\vec{\eta}|^2)$$

$$\vec{Y} = O(|\vec{\eta}|^2) \qquad . \tag{5.22}$$

We have extended the notation for the matrix of first derivatives introduced for $\vec{U}$ and $\vec{V}$ to $\hat{F}$ and $\hat{G}$. Moreover we have set — for simplicity in notation but with no loss of generality — $\vec{c} = 0$.

The Eqs. (5.20), (5.21) are of the form

$$\vec{f}(\vec{w}) = 0 \qquad , \tag{5.23}$$

where $w = \binom{U_i}{V_i}$ $(i = 0,1)$

and $$\vec{f}(\vec{w}) = L\vec{w} + \hat{f}(\vec{w}) \qquad . \tag{5.24}$$

L denoting a linear operator, and $\hat{f}$ a small quantity. One should remember also that they must be solved by keeping in mind the symmetry and periodicity requirements imposed by $\mathfrak{G}$ on $\vec{w}$.

The formal solution of (5.23), (5.24) can be obtained by setting first

$$\vec{w} = -L^{-1}\,\hat{f}(\vec{w}) \tag{5.25}$$

and then solving (5.25) by some perturbative iteration procedure.

The main difficulty with this scheme is that this might fail, because L^{-1} is an unbounded operator. Indeed inverting L involves solving equations of the form

$$(\vec{\omega}_0 \cdot \nabla_{\vec{\xi}})\, U_j(\vec{\xi}) = g_j(\vec{\xi}) \qquad , \qquad j = 0,1 \tag{5.26}$$

with the conditions that U_j and g_j are functions on $T^n(\vec{\xi} \in T^n)$, and $g_j(-\vec{\xi}) = -g_j(\vec{\xi})$, $U_j(-\vec{\xi}) = U_j(\vec{\xi})$.

Expanding $g_j(\vec{\xi})$ as a Fourier series ($\vec{m}$ denotes a vector with integer components)

$$g_j(\vec{\xi}) = \sum_{\vec{m}} g_{\vec{m}}^{(j)} e^{i \vec{m} \cdot \vec{\xi}}$$

$$g_{-\vec{m}}^{(j)} = -g_{\vec{m}}^{(j)} \quad , \quad (g_{\vec{0}} = 0) \quad , \tag{5.27}$$

Equation (5.26) is solved by

$$U_j(\vec{\xi}) = -i \sum_{\substack{\vec{m} \\ \vec{m} \neq \vec{0}}} \frac{g_{\vec{m}}^{(j)}}{\vec{m} \cdot \vec{\omega}_0} e^{i \vec{m} \cdot \vec{\xi}} = \Omega \circ g_j \tag{5.28}$$

Ω is a shorthand notation for the functional operation defined by (5.28) itself.

A condition of the type of (5.4) could guarantee that the series in (5.28) converges (in particular it forces the $\omega_{0\alpha}$'s to be incommensurate), but the unboundedness of L^{-1} implied by the presence of small denominators in (5.28) may accumulate in the iteration procedure, and cast doubts on the perturbative solution of (5.25). The difficulty can be overcome only if an iterative scheme is devised such that the error ε_r at the r-th step is made proportional to some power $\beta > 1$ of ε_{r-1}. Moser and Hald were able to produce such a scheme, with $\beta = 2$; thus ensuring the convergence of the iteration under the assumption (5.4). The procedure is a clever generalization of Newton's method. Indeed, the latter would give, for Eq. (5.23),

$$w_{r+1} = w_r - [f'(w_r)]^{-1} \cdot f(w_r) \tag{5.29}$$

where — due to (5.24) — one can start, e.g., with $w_0 = 0$, and $f'(w_r)$ is the Fréchet derivative of $f(w)$ at $w = w_r$. We have dropped the vector notation for simplicity. As one can check from (5.24), $f'(w_r)$ is close to L and its inverse may not exist just for the same reasons discussed before. Moser and Hald replaced (5.29) with a double set of recursion formulae

$$w_{r+1} = w_r - a_r\, f(w_r) \quad , \tag{5.30-a}$$

$$a_{r+1} = a_r\, [2\,\mathbb{I}_n - a_r\, f'(w_{r+1})] \quad . \tag{5.30-b}$$

Clearly for $r\to\infty$, w_r tends to the solution of (5.23), say $\tilde{w}$, as in (5.29), whereas a_r tends to $[f'(\tilde{w})]^{-1}$. The advantage of (5.30) is that $f'(w_r)$ never appears at the denominator. Moreover its convergence rate is very fast. Indeed, define

$$z_r = w_r - \tilde{w} \quad , \quad \zeta_r = \mathbb{I}_n - a_r\, f'(w_r) \quad , \tag{5.31}$$

which are quantities both tending to zero as the iteration proceeds. By a Taylor expansion, we can write

$$0 = f(\tilde{w}) = f(w_r) - f'(w_r)z_r + 0(z_r^2) \tag{5.32}$$

and using (5.30-a)

$$z_{r+1} = z_r - a_r\, [f'(w_r)z_r + 0(z_r^2)] =$$

$$= (\mathbb{I}_n - a_r\, f'(w_r))z_r + 0(z_r^2) = \zeta_r z_r + 0(z_r^2) \quad . \tag{5.33}$$

On the other hand we can write (5.30-b) in the form

$$\zeta_{r+1} \doteq \mathbb{I}_n - a_{r+1}\, f'(w_{r+1}) = [\mathbb{I}_n - a_r\, f'(w_{r+1})]^2 \quad . \tag{5.34}$$

But on making use of (5.32),

$$\mathbb{I}_n - a_r\, f'(w_{r+1}) = \zeta_r - a_r\, [f'(w_{r+1}) - f'(w_r)] =$$

$$= \zeta_r - a_r(z_{r+1} - z_r)f'' =$$

$$= \zeta_r + a_r^2\, f'(w_r)f''z_r + 0(z_r^2) \tag{5.35}$$

where the second derivative f'' is to be calculated at some intermediate point between w_r and w_{r+1}, and (5.3) used. Thus we have from (5.33) and (5.34)-(5.35), respectively

$$|z_{r+1}| \leq |z_r \zeta_r| + C_1 \, |z_r|^2 \quad ,$$
$$|\zeta_{r+1}| < C_2 \, (z_r^2 + \zeta_r^2) \quad , \tag{5.36}$$

for some finite (positive) constants C_1, C_2. Then (5.36) implies, that for some finite (positive) constant C_3,

$$\sqrt{z_{r+1}^2 + \zeta_{r+1}^2} \leq C_3 \, (z_r^2 + \zeta_r^2) \quad , \tag{5.37}$$

showing that the convergence is quadratic. In other words, starting the iteration with an error of the order ε [i.e. replacing in the right-hand side of (5.20), (5.21) the argument of $\hat{F}, \hat{G}$ and their derivatives by $(\vec{\xi}, 0)$, and dropping the products of such derivatives with U_1, V_1 respectively, so that the equations have constant coefficients], one ends up, after r steps, with an error of the order of ε^{2r}. We can now return to our original equations. As mentioned previously, they have the form (5.26) or, to be more precise

$$(\vec{\omega}_0 \cdot \nabla_{\vec{\xi}}) U_j - V_j = A_j(\vec{\xi}) \quad ,$$
$$(\vec{\omega}_0 \cdot \nabla_{\vec{\xi}}) V_j = B_j(\vec{\xi}) \quad , \qquad j = 0,1 \tag{5.38}$$

at any iteration. Using the notation of (5.26)-(5.28), the solution of (5.38) satisfying all symmetry and periodicity requirements reads finally

$$V_j = \Omega \circ B_j - A_{\vec{0}}^{(j)}$$
$$U_j = \Omega \circ (A_j - A_{\vec{0}}^{(j)}) + \Omega^2 \circ B_j \quad , \tag{5.39}$$

where $A^{(j)}_{\vec{0}}$ is the constant term in the Fourier expansion of $A_j(\vec{\xi})$. Now, for any function $g_j(\vec{\xi})$ such as that given in (5.27) (so are A_j and B_j above), one has for the Fourier coefficients an exponential decay,

$$|g^{(j)}_{\vec{m}}| \le \alpha\, e^{-\delta\|\vec{m}\|} \quad , \tag{5.40}$$

where α,δ are positive constants, and by (5.4)

$$|\Omega \circ g_j| \le \sum_{\vec{m}} \frac{|g^{(j)}_{\vec{m}}|}{|\vec{m}\cdot\vec{\omega}_0|}\, e^{\rho\|\vec{m}\|} \le$$

$$\le \frac{\alpha}{a} \sum_{\substack{\vec{m} \\ \vec{m}\neq\vec{0}}} \|\vec{m}\|^b e^{-(\delta-\rho)\|\vec{m}\|} \quad , \tag{5.41}$$

with ρ denoting the radius of convergence of $\Omega \circ g_j$. Thus, if $\rho < \delta$,

$$|\Omega \circ g_j| \le \frac{2\alpha}{a}\, \Gamma(b-1)(\delta-\rho)^{-b-n} \quad . \tag{5.42}$$

This, of course, concludes the proof, showing how the domain in $\vec{\xi}$ and $\vec{\eta}$ should be restricted in order to guarantee that the transformed equation (5.16) holds with $\vec{X}, \vec{Y}$ given by (5.22) and solutions linear in $\vec{\eta}$ (as in (5.19)). Of course this implies as well that a particular family of solutions is

$$\vec{\xi} = \vec{\omega}_0 t + \vec{\xi}(0) \quad , \quad \vec{\eta} = 0 \quad , \tag{5.43}$$

which when inserted into (5.14), gives

$$\vec{\Theta} = \vec{U}(\vec{\omega}_0 t + \vec{\xi}(0)\,,0) = \vec{\xi} + \vec{U}_0(\vec{\omega}_0 t + \vec{\xi}(0)\,,0) \quad ,$$

$$\vec{I} = \vec{V}(\vec{\omega}_0 t + \vec{\xi}(0)\,,0) = \vec{V}_0(\vec{\omega}_0 t + \vec{\xi}(0)\,,0) + \vec{c} \quad , \tag{5.44}$$

(where we have restored the factor $\vec{c}$). Comparing with (5.6) shows that in this case $\vec{\theta} = \vec{\xi}$.

6. A Few Examples of Non-Integrable Systems

If one considers the n-dimensional harmonic oscillator defined by the Hamiltonian

$$H = \frac{1}{2}\sum_{\alpha=1}^{n}\left(\frac{p_\alpha^2}{m_\alpha} + m_\alpha\omega_\alpha^2 q_\alpha^2\right) \quad , \tag{6.1}$$

the canonical transformation defined by

$$q_\alpha = \sqrt{\left(\frac{2I_\alpha}{\omega_\alpha m_\alpha}\right)}\cos\Theta_\alpha$$

$$p_\alpha = \sqrt{\left(2\omega_\alpha I_\alpha m_\alpha\right)}\sin\Theta_\alpha \quad , \quad \alpha = 1,\ldots,n \tag{6.2}$$

changes it into action-angle variables. The new Hamiltonian is

$$H' = H'(\{I_\alpha\}) = \sum_{\alpha=1}^{n}\omega_\alpha I_\alpha \quad . \tag{6.3}$$

Thus any integrable system can be expected to be reducible, by a suitable coordinate transformation (in general non-linear), to the form (6.1). The corresponding phase-portrait is then a direct sum of n one-degree of freedom harmonic-oscillator-like phase portraits of the form illustrated in Fig. 6. Notice that the system takes an infinitely long time to go, along a separatrix, from an hyperbolic point to the next. As was discussed before, it is not so if the system is perturbed with a non-integrable interaction εH_1.

Let an integrable system be described, in the canonical coordinates $(\pi_\alpha, \chi_\alpha)$, $\alpha = 1,\ldots,n$ by the real analytic Hamiltonian $H = H(\{\pi_\alpha, \chi_\alpha\})$. Introduce the complex coordinates

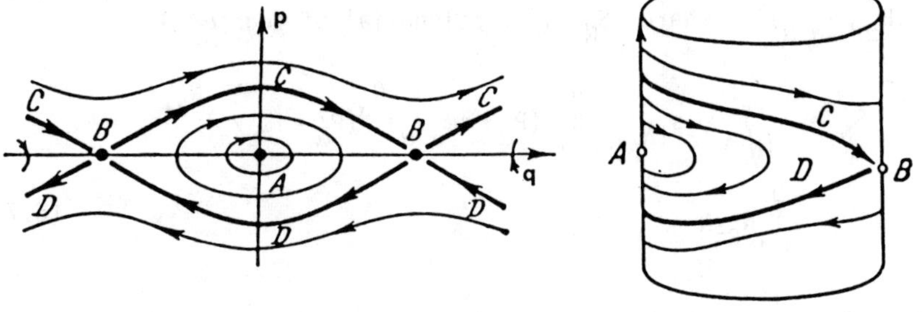

Fig. 6

$$z = \pi_\alpha + i\ \omega_\alpha\chi_\alpha \qquad ; \qquad \alpha = 1,\ldots,n \quad . \tag{6.4}$$

There exists a canonical coordinate transformation from $\{(\pi_\alpha, \chi_\alpha)\}$ to, say $\{(P_\alpha, Q_\alpha)\}$ such that the Hamiltonian is a function of $\tau_\alpha = \frac{1}{2}(P^2 + Q^2)$, $\alpha = 1,\ldots,n$ at least up to degree M: $H' = H_B(\{\tau_\alpha\}) + R$, and the remainder $R = 0\,(|P_\alpha| + |Q_\alpha|)^{m+1}$, where m is the minimum order of resonance of the ω_α's. Recall that the frequencies $\{\omega_\alpha\}$ are said to be resonant if there exists a set of integers $\{M_\alpha, \ \alpha = 1,\ldots,n\}$ such that

$$\sum_{\alpha=1}^{n} M_\alpha\omega_\alpha = 0 \quad . \tag{6.5}$$

The number $M = \sum_{\alpha=1}^{n} M_\alpha$ is referred to as the order of resonance. Indeed, writing H in terms of z_α's and $\bar{z}_\alpha$, the generic monomial term of degree N in it reads

$$z_1^{N_1}\ldots z_\alpha^{N_\alpha}\ldots z_n^{N_n}\ \bar{z}_1^{N_{n+1}}\ldots\bar{z}_\alpha^{N_{n+\alpha}}\ldots\bar{z}_n^{N_{2n}} \quad , \quad \sum_{\beta=1}^{2n} N_\beta = N \quad . \tag{6.6}$$

Perform now the transformation of generating function $w_N = \sum_{\alpha=1}^{n} P_\alpha \omega_\alpha \chi_\alpha + S_N(\{P_\alpha, \omega_\alpha\chi_\alpha\})$, where S_N is a polynomial of degree N,

$$S_N = \sum_{\substack{\{N_\beta\} \\ \sum_{\beta=1}^{2n} N_\beta \le N}} s_{\{N_\beta\}} \prod_{\alpha=1}^{n} (P_\alpha + i\omega_\alpha\chi_\alpha)^{N_\alpha} (P_\alpha - i\omega_\alpha\chi_\alpha)^{N_{k+\alpha}} \quad . \tag{6.7}$$

A transformation of this sort obviously changes, in the Taylor expansion of H, only terms of degree $\geq N$. In particular, for Eq. (6.6), the coefficient of the term of degree N is proportional to

$$\underset{\sum_{\beta=1}^{2n} N_\beta = N}{s_{\{N_\beta\}}} \cdot \sum_{\alpha=1}^{n} \omega_\alpha (N_{n+\alpha} - N_\alpha) \quad . \tag{6.8}$$

If $N \leq M$, this is in general different from zero: it is zero only if $N_{n+\alpha} = N_\alpha$, $\forall\alpha = 1,\ldots,n$. Selecting only the non-zero coefficients $s_{\{N_\beta\}}$ corresponding to the latter condition gives a generating function whose monomial terms of Eq. (6.6) are functions of $|z_\alpha|^2 = \pi_\alpha^2 + \omega_\alpha^2 \chi_\alpha^2$. These in turn transform into functions of τ_α (because then S_N depends only on $(P_\alpha^2 + \omega_\alpha^2 \chi_\alpha^2)$ $\alpha = 1,\ldots,n$).

Transforming now the original system by a sequence of transformations generated by $w_2, w_3, \ldots, w_M$ leads to a point where — due to the assumed resonance condition — (6.8) can be zero whatever the choice of $s_{\{N_\beta\}}$. H_B is called Birkhoff's normal form of the Hamiltonian H. Since now

$$\frac{\partial H_B}{\partial P_\alpha} = \frac{\partial H_B}{\partial \tau_\alpha} P_\alpha \quad , \quad \frac{\partial H_B}{\partial Q_\alpha} = \frac{\partial H_B}{\partial \tau_\alpha} Q_\alpha \quad , \quad \alpha = 1,\ldots,n \tag{6.9}$$

one has, neglecting R,

$$\dot{Q}_\alpha = \frac{\partial H'}{\partial P_\alpha} \cong \frac{\partial H_B}{\partial \tau_\alpha} P_\alpha$$

$$\dot{P}_\alpha = - \frac{\partial H'}{\partial Q_\alpha} \cong - \frac{\partial H_B}{\partial \tau_\alpha} Q_\alpha \quad , \quad \alpha = 1,\dots,n \tag{6.10}$$

whence

$$\frac{d}{dt}(P_\alpha^2 + Q_\alpha^2) = 2\dot{\tau}_\alpha \cong 0 \quad , \quad \alpha = 1,\dots,n \tag{6.11}$$

and the I_α's are well approximated by functions of $\{\tau_\alpha\}$ for small oscillations around $\{Q_\alpha = 0,\ P_\alpha = 0,\ \alpha = 1,\dots,n\}$.

As a first example of a non-integrable system, we discuss a non-integrable system with one degree of freedom $(n = 1)$, which in the coordinates $z, \bar{z}$ (cf Eq. (6.4)) reads

$$H = \frac{1}{2}\omega|z|^2 + 2i \sum_{\substack{M=-\infty \\ M\neq 0}}^{+\infty} \sum_{\substack{N,L \\ N+L=3}} h_{NL}^{(M)} z^N \bar{z}^L e^{iM\omega_0 t} + O(z^3) \tag{6.12}$$

where the frequency $\omega = (\frac{1}{3} + \varepsilon)\omega_0$. Reducing H to normal form, one obtains a Hamiltonian in which all the third-degree terms are eliminated, but the "resonant" ones, i.e. those for which

$$\omega(N - L) + M\omega_0 = 0 \quad , \tag{6.13}$$

remain. They correspond to $\varepsilon = 0$, and $N - L + 3M = 0$ (with $N + L = 3$), namely to $N = 0$, $L = 3$, $M = 1$ and $N = 3$, $L = 0$, $M = -1$. Thus H can be written

$$H = \frac{1}{2}\omega|z|^2 + 2i\,(hz^3 e^{-i\omega_0 t} - \bar{h}\bar{z}^3 e^{i\omega_0 t}) \tag{6.14}$$

where the coefficients are h, $-\bar{h}$ so as to guarantee reality of H.

By the time-dependent coordinate transformation $\zeta = ze^{-i\frac{1}{3}\omega_0 t}$, one has finally the time-independent Hamiltonian

$$H_0 = \frac{1}{2}\omega\,|\zeta|^2 + 2i(h\zeta^3 - \bar{h}\bar{\zeta}^3) \quad . \tag{6.15}$$

Making now a last transformation to real coordinates, $\zeta = x + iy$ and assuming, for simplicity, $h = -\frac{1}{12}k$ real,

$$H_0 = \frac{1}{2}\omega\left(x^2 + y^2\right) + k\left(x^2 y - \frac{1}{3}y^3\right) \quad . \tag{6.16}$$

Hénon and Heiles studied in detail the Poincaré map for a dynamical system obtained from (6.16) by doubling the number of degrees of freedom with the addition of two momenta p_x, p_y respectively conjugate to x and y, and assuming H_0 to be the potential energy. By a suitable rescaling of the constants:

$$H_H = \frac{1}{2}(p_x^2 + p_y^2) + \frac{1}{2}(x^2 + y^2) + x^2 y - \frac{1}{3}y^3 \quad . \tag{6.16'}$$

The corresponding equations of motion are

$$\ddot{x} = -x - 2xy \quad ,$$

$$\ddot{y} = -y + y^2 - x^2 \quad . \tag{6.17}$$

The energy surface $H = \mathscr{E}$ can be easily seen to be non-compact for $\mathscr{E} > \frac{1}{6}$, therefore Eqs. (6.17) were studied for a set of choices of in the range $0 < \mathscr{E} \leq \frac{1}{6}$. Two different methods of approach were utilized: one analytic (the perturbative construction of a second constant of the motion besides H_H, say $J = J(x, y, p_x, p_y) = c$ which, if analytic, would exhibit the integrability of the system), the other numerical (one computer-calculated orbit of (6.17) is intersected with S_y). The images of the numerical Poincaré map were then compared with $J \cap \mathscr{S}_y$, with J characterized by the initial conditions of the orbit selected. The perturbative calculation was carried out up to the 7-th order. The results are shown in Fig. 7.

One can easily see how, up to $\mathscr{E} = 1/2\, \mathscr{E}_r (\mathscr{E}_r = 1/6)$ there is a global agreement between the analytical and numerical results (even though progressively worsening with increasing $\mathscr{E}$); and one can recognize in the latter a clear evidence of the existence of invariant curves, which are but the intersections with $\mathscr{S}_y$ of the conserved tori. However as $\mathscr{E}$ is further increased, there appears at first clear evidence that some of the tori are disrupted and others badly deformed, and finally, very close to $\mathscr{E} = \mathscr{E}_r$, the phase space is almost filled up, mostly in a stochastic way (one should not forget that the picture shows the intersections of $\mathscr{S}_y$ with a <u>single</u> orbit).

It is very instructive to compare the above results with those of a similar analysis performed on the dynamical system (Toda-like) characterized by:

$$H_T = \frac{1}{2}(p_x^2 + p_y^2) + \frac{1}{12}\left[e^{2y}\cosh(2\sqrt{3}\,x) + \frac{1}{2}e^{-4y}\right] - \frac{1}{8} \tag{6.18}$$

because it coincides with (6.16') if terms of order higher than the third are truncated in the Taylor expansion; moreover it is integrable. Figure 8 shows the iso-energy curves for the potential part (6.16) of (6.16') and (6.18). Besides H_T, Eq. (6.18) has a second conserved quantity,

$$J = \frac{2}{3}p_x(p_x^2 - 3p_y^2) + \frac{1}{6}p_x\left[e^{2y}\cosh(2\sqrt{3}\,x) - e^{-4y}\right] -$$

$$- \frac{1}{2\sqrt{3}}p_y\, e^{2y}\sinh(2\sqrt{3}\,x) \tag{6.19}$$

and the phase portrait in $\mathscr{S}_y$ can be obtained by eliminating x and p_x between (6.18) and (6.19). The version of the latter truncated after third-order terms reads

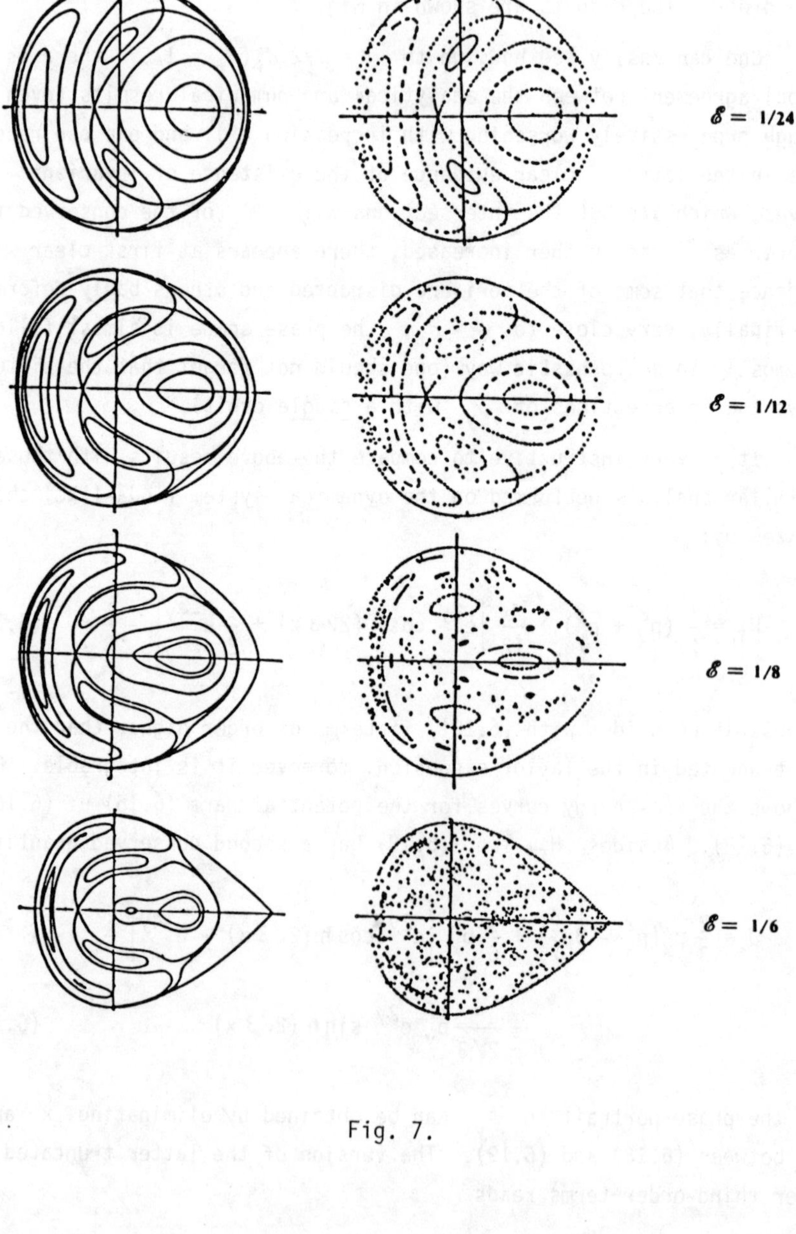

Fig. 7.

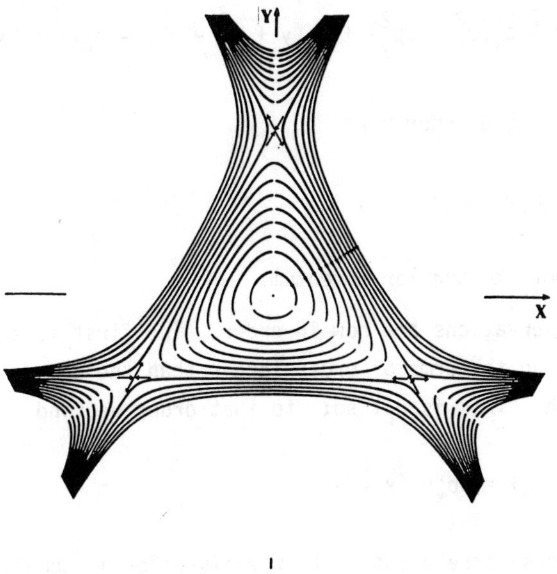

Fig. 8.a.

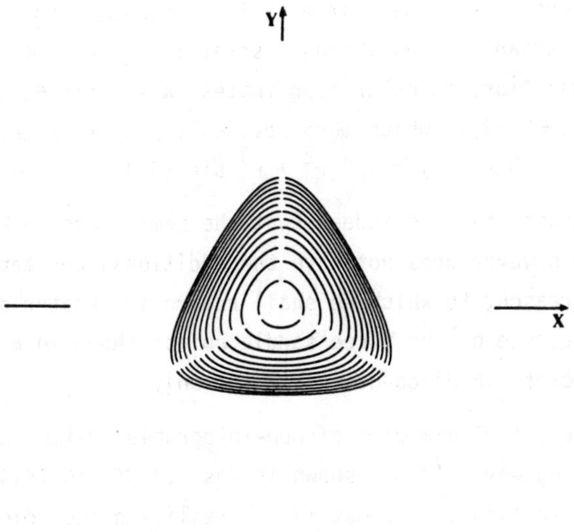

Fig. 8.b.

$$J^{(3)} = \frac{2}{3} p_x(p_x^2 - 3p_y^2) + p_x(y + x^2 - y^2) - p_y(x + 2xy) \quad , \tag{6.20}$$

whereas its second-order part

$$J^{(2)} = p_x y - p_y x \quad , \tag{6.21}$$

is nothing but the angular momentum.

Two observations are now in order. The first is that one might expect that, up to terms of order larger than the third, $\{J^{(3)}, H_H\}$ should vanish. This is not so; to that order we find

$$\{J^{(3)}, H_H\} = 4p_x p_y(x - y) \quad . \tag{6.22}$$

(It is of course irrelevant that it vanishes on a submanifold of $\mathscr{S}_y$.) The second is that the existence of the conserved quantity J, which is clearly an "extension" of angular momentum, could be expected from Noether's theorem since the system (6.18) exhibits a configurational, rotational invariance, even though discrete: $(x, y) \to (-1/2\,(x + \sqrt{3}\,y),\ 1/2\,(\sqrt{3}\,x - y))$ [or, in polar coordinates $x = r\cos\theta$, $y = r\sin\theta$: $(r, \theta) \to (r, \theta + 2\pi/3)$, which more obviously preserves the potential part of (6.17): $H_0(r, \theta) = 1/2\, r^2 + r^3 \sin(3\theta)$].

A disturbing fact is indeed that the same symmetry holds also for (6.17), which however does not have the additional constant of the motion. The reason, to which we shall return in greater detail later on, is the presence of the three saddle points shown in Fig. 8.a characteristic of the Hénon-Heiles system only.

A second set of examples of non-integrable systems can be derived in the following way. It was shown in Eqs. (4.34) to (4.48) that the explicit construction of the matrix $\mathscr{T}$ realizing the Poincaré mapping is quite laborious even in the simple case of 2 degrees of freedom. For non-integrable systems, it cannot be expected to be realizable exactly: at best one can hope to be able to find, in the neighbourhood of the tori whose existence is guaranteed by the KAM theorem when the non-integrable perturbation is small, a perturbative description of the mapping. On the other hand to have an explicit form of $\mathscr{T}$ available is unquestionably the most convenient way to produce an accurate phase portrait of the system. The latter is in turn a very effective tool to realize whether or not the system is likely to be integrable, and — in case it isn't — which are the regions of phase space where its behaviour is most manifestly stochastic.

Thus there is an intrinsic interest in studying dynamical systems for which the matrix $\mathscr{T}$ can be explicitly derived. A class of such systems is characterized by periodic, time-dependent Hamiltonians of the following form

$$H = H(\{q_\alpha , p_\alpha\} ; t) = \begin{cases} \sum_{\alpha=1}^{n} \frac{p_\alpha^2}{2m_\alpha} & 0 \le t < T - \eta \\ V(\{q_\alpha\}) & T - \eta \le t < T \; ; \; \eta \ll T \end{cases} \tag{6.23}$$

$$H(\{q_\alpha , p_\alpha\} ; t + T) \equiv H(\{q_\alpha , p_\alpha\} ; t) \quad . \tag{6.24}$$

In other words it is required that the force

$$\vec{F} \equiv \left\{ F_\alpha = - \frac{\partial V}{\partial q_\alpha} \quad ; \quad \alpha = 1,\ldots,n \right\} \tag{6.25}$$

act impulsively and that during the short (possibly vanishing) time η in which it is switched on, the kinetic energy might be neglected with respect to the potential energy V.

Denoting simply by (q , p) the coordinates of the selected phase plane $\mathscr{S}$, the Hamiltonian defined by (6.23), (6.24) gives right away for their successive images the following discrete map

$$q^{(k+1)} = q^{(k)} + (T - \eta)p^{(k)}$$

$$p^{(k+1)} = p^{(k)} - \eta \frac{\partial V}{\partial q}(q^{(k+1)}) \quad , \quad k \geq 0 \quad . \qquad (6.26)$$

The superscript index k refers to the order of iteration of $\mathscr{T}$. The initial condition $(q^{(0)}, p^{(0)})$ is now essentially arbitrary, because the remaining variables are supposed to adjust in such a way that the energy is kept constant (if required). Equations (6.26) can be reduced to a single equation, e.g. by eliminating $p^{(k)}$,

$$q^{(k+2)} - 2q^{(k+1)} + q^{(k)} = - \eta(T - \eta) \frac{\partial V}{\partial q}(q^{(k+1)}) \quad , \; k \geq 0 \qquad (6.27)$$

in which one may easily recognize a discrete version of Newton's equation. If η is negligible with respect to T, periodicity may be built in, by replacing (6.23) with

$$H = \frac{1}{2} \sum_{\alpha=1}^{n} \left(\frac{p_\alpha^2}{m_\alpha} + m_\alpha \Omega^2 q_\alpha^2 \right) + \delta(t - MT)\, V(\{q_\alpha\}) \quad , \quad M \in \mathbb{Z} \qquad (6.28)$$

where δ is Dirac's delta function. The equivalents of (6.26) and (6.27) in this case read respectively,

$$q^{(k+1)} = c\, q^{(k)} + \frac{s}{m\Omega} p^{(k)} \quad ,$$

$$p^{(k+1)} = - m\Omega s\, q^{(k)} + c\, p^{(k)} - \frac{\partial V}{\partial q}(q^{(k+1)}) \quad , \quad k \geq 0 \qquad (6.29)$$

where

$$c = \cos(\Omega T) \quad , \quad s = \sin(\Omega T) \quad , \qquad (6.30)$$

and

$$q^{(k+2)} - 2q^{(k+1)} + q^{(k)} = 2(c-1)q^{(k+1)} - \frac{s}{m\Omega}\frac{\partial V}{\partial q}(q^{(k+1)}) \quad ,$$

$$k \geq 0 \quad . \qquad (6.31)$$

Both (6.29) and (6.30) can be obtained by direct integration of the equations of motion over the time interval T, which separates two successive crossings of $\mathscr{S}$.

In the sequel a few examples based on the above scheme will be discussed. They will be selected from different fields of physics, in order to exhibit the variety of applications that the method can have.

The first example is that of colliding-beam machines in high energy experiments. These are essentially betatron-like machines which are operated in pairs and act as storage rings (one for each type of particle e.g., protons and antiprotons, or e^+ and e^-). After every revolution the two beams are made to collide with each other over a very small region along the central trajectory (the two machines are mutually tangential). The particle coordinates relative to the latter are thus transformed by a transformation which is of the form (6.29), where one should think of time as being replaced by an angular coordinate which is the betatron phase ψ along the central trajectory, and where $\frac{p}{m}$ should be replaced by $\dot{q} \doteq \frac{dq}{d\psi}$. Several choices of the driving term $-\frac{\partial V}{\partial q}$ have been used. The most realistic is of the form

$$F(q) = -\frac{\partial V}{\partial q} = B\,\frac{1 - \exp\left(-\dfrac{q^2}{2\sigma^2}\right)}{q} \qquad (6.32)$$

where B is a constant, obviously proportional to the average beam current, and σ is a Gaussian standard deviation describing the particle distribution around the central trajectory.

The results of numerical iteration of (6.29) are shown in Fig. 9. One should notice how the phase portrait — which with the coarse-grained scale of the upper picture seems to exhibit a smooth collection of invariant tori — indeed shows regions where the distribution of representative points of the orbit is quite random.

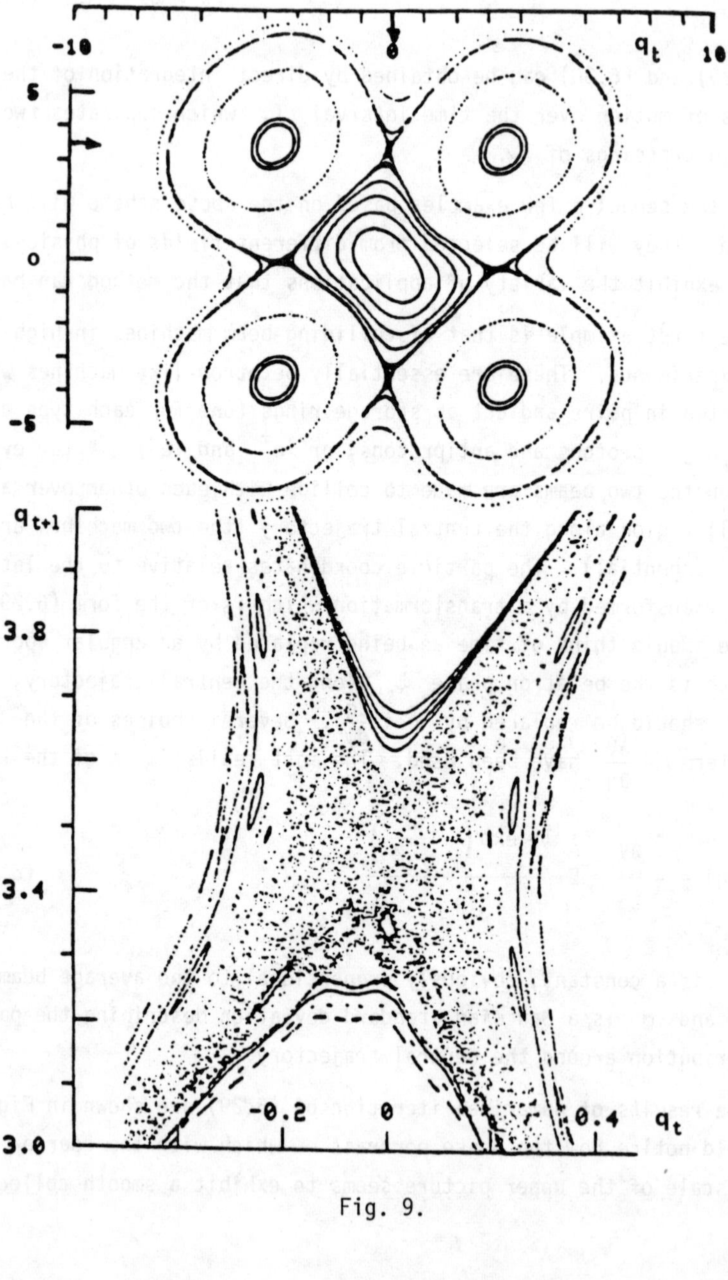

Fig. 9.

Another choice, which is meant to describe not the collision experiment but the steady situation of each beam; devised to study the effect of spatial inhomogeneities of the magnetic field, is

$$F(q) = Aq^2 \tag{6.33}$$

where A is a suitable constant. The corresponding phase portrait is similar to that of the previous example, except for the appearance of a resonance of order five instead of four, due to the different parity of the function in (6.33) with respect to that in (6.32). The results are shown in Fig. 10.

The existence of invariant tori in the phase space of the perturbed system implies that for most of the initial conditions, the motion of a system close enough to an integrable one remains quasi-periodic with a finite maximum number of frequencies. It is natural to inquire about what happens of the remaining phase curves, whose initial conditions lie in the gaps between those tori which are preserved under perturbation. In order to do this, let us study in detail the map (6.31) corresponding to Fig. 10, that writes

$$q^{(k+1)} + q^{(k-1)} = 2c\, q^{(k)} + A'[q^{(k)}]^2 \quad , \qquad k \geq 1 \tag{6.34}$$

where $A' = sA/m\Omega$ is a constant, in more detail. There are two fixed points of (6.34) in the phase plane $(q^{(k)}, q^{(k+1)})$, say P_1, P_2 of coordinates, respectively

$$\hat{q}_1^{(0)} = \hat{q}_1^{(1)} = 0 \quad , \quad \hat{q}_2^{(0)} = \hat{q}_2^{(1)} = \frac{2}{A'}(1-c) \quad . \tag{6.35}$$

Let us first determine the nature of such fixed points by linearizing (6.34) around them, as discussed in (4.45) and successive equations. Upon setting in (6.34)

$$q^{(k)} = \hat{q}^{(k)} + \eta^{(k)} \quad , \quad k \geq 0 \tag{6.36}$$

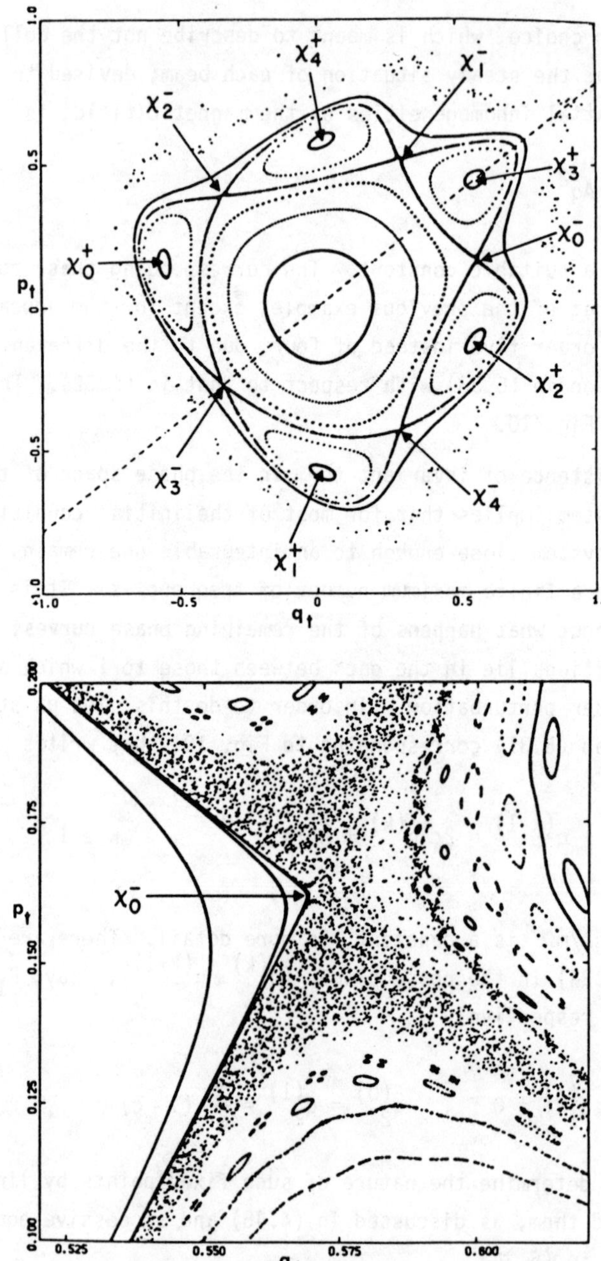

χ_4^+
χ_1^-
χ_2^-
χ_3^+
χ_0^-
χ_0^+
χ_2^+
χ_3^-
χ_4^-
χ_1^+
p_t
q_t
1.0
0.5
0
-0.5
-1.0
-0.5
0.5
1.0
χ_0^-
p_t
q_t
0.200
0.175
0.150
0.125
0.100
0.525
0.550
0.575
0.600

Fig. 10.

where $\hat{q}^{(k)} \doteq \hat{q}_i^{(k,\, \mathrm{mod}\, 2)}$, $i = 1$, or 2; and keeping only terms linear in $\eta^{(k)}$, one finds

$$\eta^{(k+1)} + \eta^{(k-1)} = 2(c + A'\hat{q}^{(k)})\eta^{(k)} \quad , \quad k \geq 1 \tag{6.37}$$

which reads, for $i = 1$

$$\eta^{(k+1)} + \eta^{(k-1)} = 2c\,\eta^{(k)} \quad , \quad k \geq 1 \tag{6.38}$$

and for $i = 2$

$$\eta^{(k+1)} + \eta^{(k-1)} = 2(2-c)\,\eta^{(k)} \quad , \quad k \geq 1 \quad . \tag{6.39}$$

The corresponding 2×2 matrices for $\mathcal{T}$:

$$\begin{pmatrix} \eta^{(k+1)} \\ \eta^{(k)} \end{pmatrix} = \mathcal{T} \begin{pmatrix} \eta^{(k)} \\ \eta^{(k-1)} \end{pmatrix} \quad , \quad k \geq 1$$

are

$$\mathcal{T}_1 = \begin{pmatrix} 2c & -1 \\ 1 & 0 \end{pmatrix} \quad ; \quad \mathcal{T}_2 = \begin{pmatrix} 2(2-c) & -1 \\ 1 & 0 \end{pmatrix} \quad . \tag{6.40}$$

Equations (6.40) manifestly show that $\mathcal{T}$ is area-preserving ($\det \mathcal{T} = 1$). The eigenvalues of $\mathcal{T}_1$ are $\lambda_1 = e^{\pm i\phi}$ with $\phi = T\Omega$; thus P_1 is an elliptic point, whereas those of $\mathcal{T}_2$ are real; $\lambda_2 = (2-c) \pm \sqrt{(2-c)^2 - 1} = \lambda_2^{(\pm)}$, $\lambda_2^{(-)} = \frac{1}{\lambda_2^{(+)}} > 1$, and P_2 is a hyperbolic point. Denoting the unnormalized eigenvectors of $\mathcal{T}_2$ by $\vec{e}_+ \doteq (\lambda_2^{(+)}, 1)$, $\vec{e}_- \doteq (\lambda_2^{(-)}, 1)$ one might expect the separatrix to behave as shown in Fig. 11. However, it is not so. If one considers the set of points approaching P_2 along the separatrix in the direction $\vec{e}_+$ and those

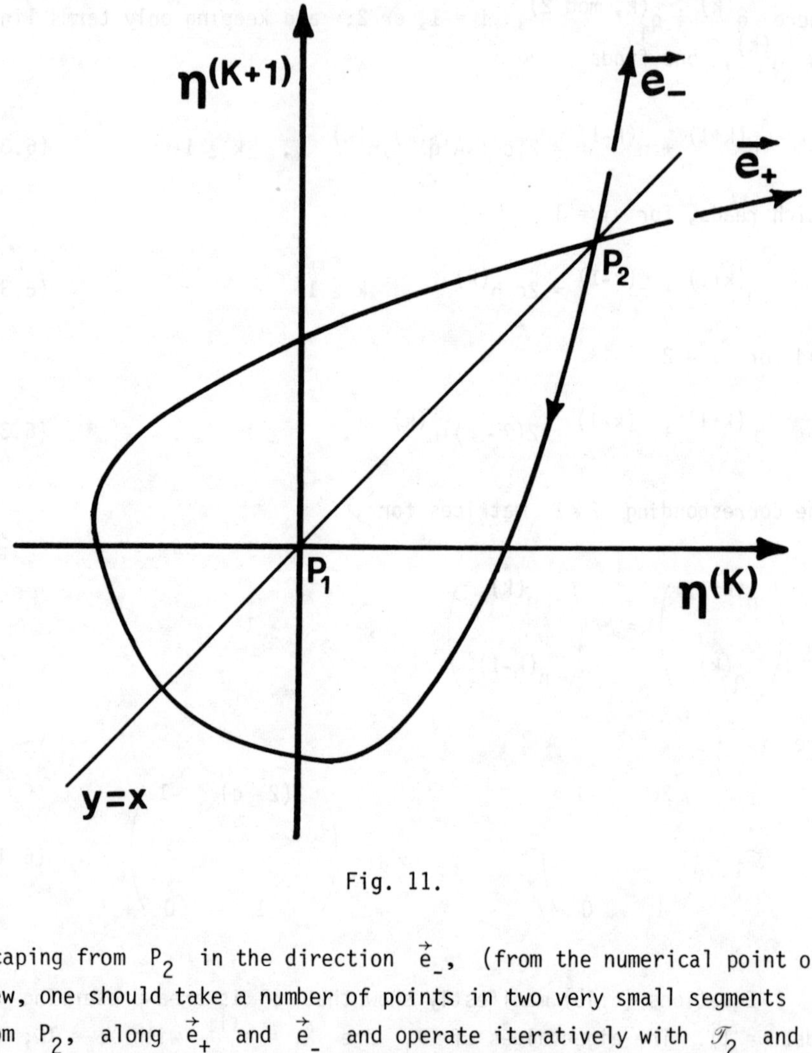

Fig. 11.

escaping from P_2 in the direction $\vec{e}_-$, (from the numerical point of view, one should take a number of points in two very small segments from P_2, along $\vec{e}_+$ and $\vec{e}_-$ and operate iteratively with $\mathcal{T}_2$ and $\mathcal{T}_2^{-1}$ respectively) one notices that the resulting curves W^+ and W^- — that we refer to as stable and unstable separatrix — do not merge in any simple way into a unique regular orbit (Fig. 12).

Indeed, it can be shown rigorously that the two curves are not the two branches of a unique smooth curve. Let us restrict, for

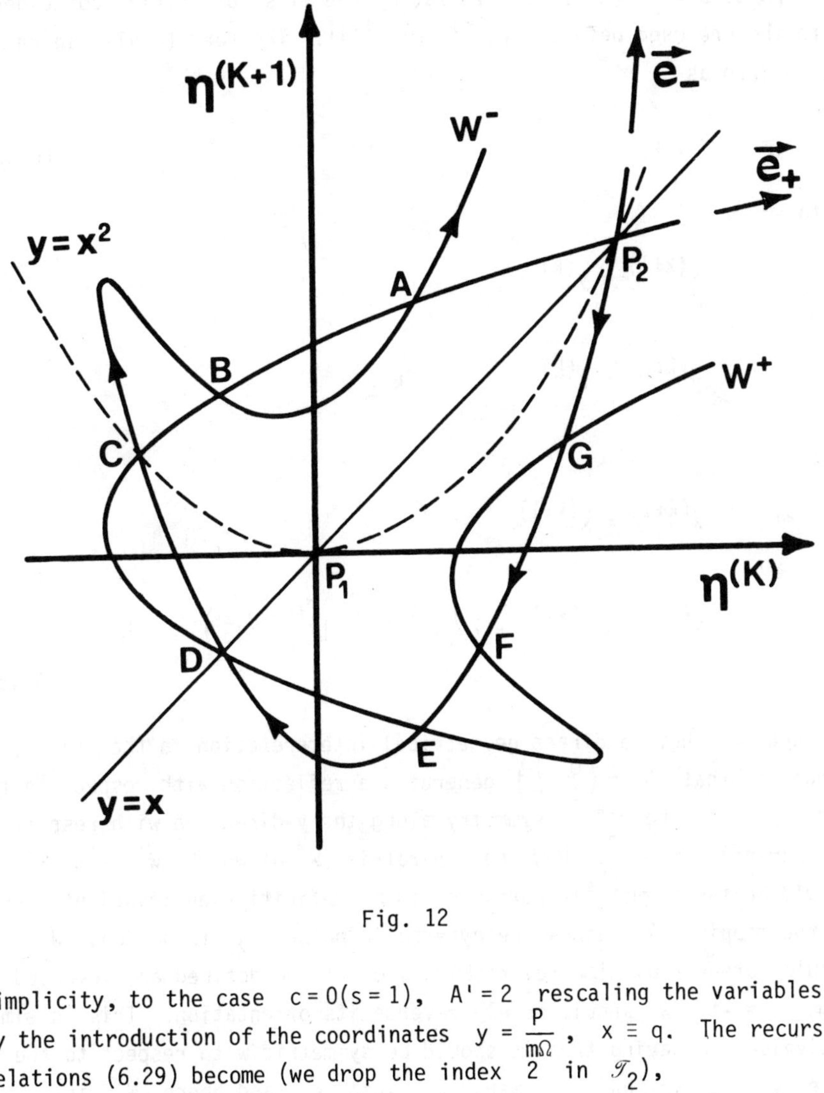

Fig. 12

simplicity, to the case $c=0(s=1)$, $A'=2$ rescaling the variables by the introduction of the coordinates $y = \frac{P}{m\Omega}$, $x \equiv q$. The recursion relations (6.29) become (we drop the index 2 in $\mathcal{T}_2$),

$$\mathcal{T}: \quad \begin{aligned} x^{(k+1)} &= y^{(k)} \\ y^{(k+1)} &= -x^{(k)} + 2[y^{(k)}]^2 \quad , \quad k \geq 0 \quad . \end{aligned} \tag{6.41}$$

[The phase plane $(x^{(k)}, y^{(k)})$ is, by the first of (6.41), coincident with the one used before, $(q^{(k)}, q^{(k+1)})$]. The form (6.41) can be factorized as

$$\mathcal{T} = \mathcal{T}'' \circ \mathcal{T}' \tag{6.42}$$

with

$$\mathcal{T}' : \begin{array}{l} \tilde{x}^{(k+1)} = y^{(k)} \\ \tilde{y}^{(k+1)} = x^{(k)} \quad , \quad k \geq 0 \end{array}$$

and

$$\mathcal{T}'' : \begin{array}{l} x^{(k+1)} = \tilde{x}^{(k+1)} \\ y^{(k+1)} = \tilde{y}^{(k+1)} - 2\left(\tilde{y}^{(k+1)} - [\tilde{x}^{(k+1)}]^2\right) \quad , \quad k \geq 0 \quad . \end{array} \tag{6.43}$$

$\mathcal{T}'$ and $\mathcal{T}''$ have a direct geometrical interpretation in the (x, y) plane, in that $\mathcal{T}' = \begin{pmatrix} 0 & 1 \\ 1 & 0 \end{pmatrix}$ generates a reflection with respect to the line $y = x$, and $\mathcal{T}''$ a symmetry along the y-direction with respect to the parabola $y = x^2$. Now the separatrix W of which W^+ and W^- should be the asymptotic portions, is by definition an invariant curve of the mapping $\mathcal{T}$. Since the hyperbolic point P_2 is unique, W should moreover be time reversible, i.e. if one defined an "inverted" time $\tau = -t$, W should merely reverse its orientation. This is simply equivalent to saying that it should be symmetric with respect to the line $y = x$, in that $\mathcal{T}'$ maps $\vec{e}_+$ into $\vec{e}_-$ and hence W^+ into W^- (and vice-versa, since $(\mathcal{T}')^2 \equiv \mathbb{I}_2$). Thus W should be an invariant curve (modulo orientation) also of the mapping $\mathcal{T}'$. Then, due to (6.42), W should be an invariant curve of the mapping $\mathcal{T}''$ as well.

In other words if W were unique, such as in the hypothetical case shown in Fig. 11, it should be separately invariant under the two components in which the mapping $\mathscr{T}$ has been factorized, i.e. it should be symmetric with respect to reflections through the bisecting line of the phase plane and the parabola $y = x^2$ (along the y direction).

Now, for any point, say $Q \equiv (x, y)$, on a curve W with the latter symmetry, the slope of the curve at Q, say d_Q, and the slope $d_{Q''}$ of W at the image point of Q under $\mathscr{T}''$, $Q'' \doteq \mathscr{T}''Q \equiv (x, 2x^2 - y)$, will obviously satisfy the relation

$$d_Q + d_{Q''} = 4x \quad . \tag{6.44}$$

On the other hand, as shown before, $\mathscr{T}$ has a second fixed point P_1, which is elliptic. W should go around P_1. But P_1 is the origin (see (6.35)), and this implies that W will cross the parabola $y = x^2$ at another point, say M. M is obviously a fixed point of $\mathscr{T}''$, $M'' \equiv M$, and should simply have $d_{M''} = d_M$ or, by (6.44), $\left.\frac{dy}{dx}\right|_M = 2x$, i.e. $y = x^2$. In other words, the separatrix should coincide with the parabola or at least be tangent to it. However this is clearly a contradiction in the first case because a separatrix could never end up in an elliptic point; in the second one should expect a new intersection on which the whole argument could be repeated (eventually leading to the same contradictory conclusion that the separatrix should pass through P_1 in order to return to the starting point).

Thus we can conclude that W^+ and W^- are not two branches of the same curve W; they are two separate curves which can never merge smoothly. Notice now that the number of intersections between W^+ and W^- is indeed infinite. M should belong to both W^+ and W^-, and by (6.44), if it is not the origin (as it cannot be), it has an image M' with respect to $\mathscr{T}'$. $\mathscr{T}'$ is idempotent, namely $(\mathscr{T}')^{-1} \equiv \mathscr{T}'$, hence M' also belongs to both curves. On the other hand M' has an image with respect to $\mathscr{T}''$, say N', which, by the same token must belong to both W^+ and W^-, and so on indefinitely (see Fig. 12) as the hyperbolic point along the separatrix should be approached in an infinite by long time.

Now consider any loop Γ formed by the two arcs of W^+ and W^- included between two successive intersections, and compute the area of the surface σ which has Γ as its boundary,

$$A(\sigma) = \iint_\sigma dx^{(k)} dy^{(k)} \quad . \tag{6.45}$$

σ is mapped by $\mathcal{T}$ onto another surface $\mathcal{T} \circ \sigma$ with the same characteristics and we want to compute the corresponding area

$$A(\mathcal{T} \circ \sigma) = \iint_{\mathcal{T} \circ \sigma} dx^{(k+1)} dy^{(k+1)} \quad . \tag{6.46}$$

The integral in (6.46) may be computed by performing the change of variables $(x^{(k+1)}, y^{(k+1)}) \to (x^{(k)}, y^{(k)})$ which is actually realized by the matrix $\mathcal{T}$. One finds

$$A(\mathcal{T} \circ \sigma) = \iint_\sigma J \, dx^{(k)} dy^{(k)} \tag{6.47}$$

where J is the Jacobian of the above transformation. Now $J = \det \mathcal{T} = 1$, and (6.47) shows that

$$A(\mathcal{T} \circ \sigma) = A(\sigma) \tag{6.48}$$

and thus, for any integer k, $A(\mathcal{T}^k \circ \sigma) = A(\sigma) = \text{constant}$. Since the intersections get closer and closer as P_2 is approached either way [take k positive if moving along W^+, negative along W^-, since the points (see (4.56)) move as $|\lambda_2^{(+)}|^k$ and $|\lambda_2^{(-)}|^{-k}$ respectively, and we selected $\lambda_2^{(-)} > 1$, $\lambda_2^{(+)} < 1$], it follows that the loops' length becomes greater and the pattern of intersections of W^+ and W^- becomes more complicated, as Fig. 13 shows schematically ($P_2 \equiv H$).

To all effects the distribution of the arcs (stable or unstable) of the separatrix — moving in opposite directions yet being arbitrarily close — soon becomes chaotic and one finds the invariant curves in the

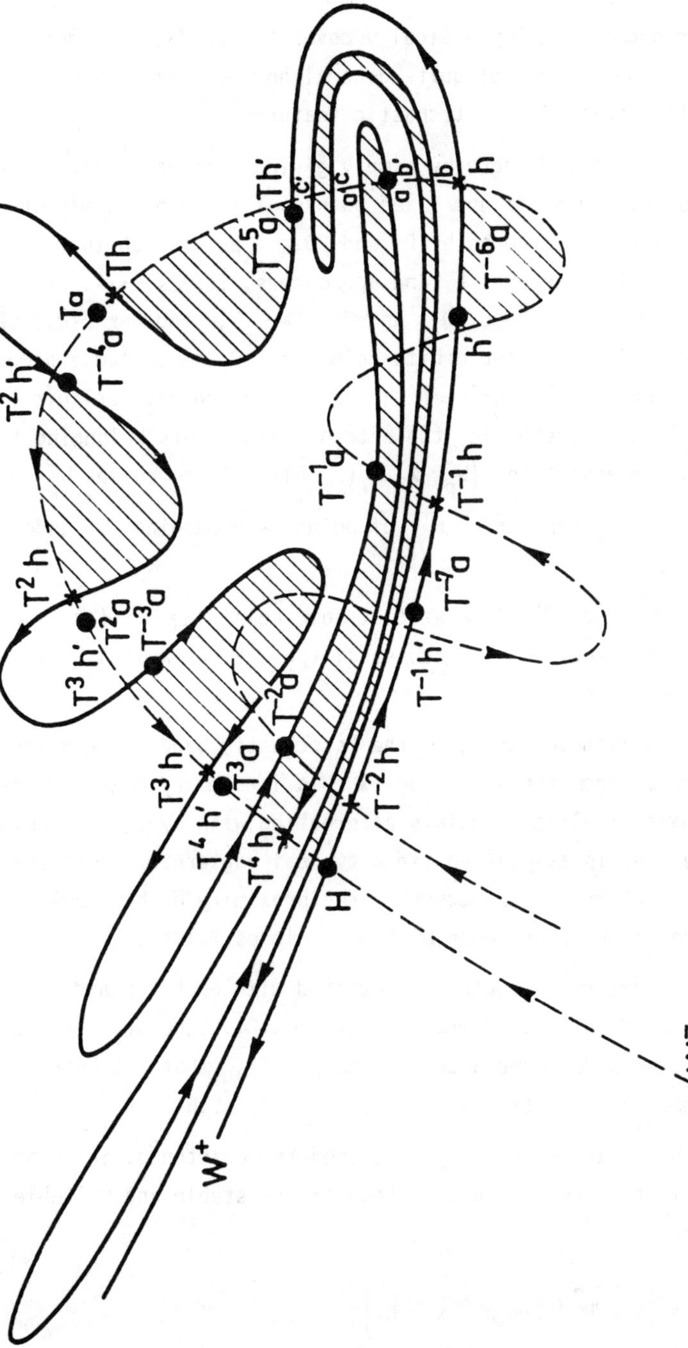
W+
W−
H
h
h'
Th
Th'
Ta
T²h
T²h'
T²a
T³h
T³h'
T³a
T⁴h
T⁴h'
T⁻¹h
T⁻¹h'
T⁻¹a
T⁻²h
T⁻²a
T⁻³a
T⁻⁴a
T⁻⁵a
T⁻⁶a
a
b
c
a'
b'
c'

Fig. 13.

neighbourhood of P_2, practically cover the surface $\mathscr{S}$ densely. This situation is in fact quite general and we shall discuss in the next section some of its stochastic features.

Let us conclude here by simply giving some necessary definitions. Let $\hat{P}, \hat{Q}$ be two different fixed points of the generic mapping $\mathscr{T}$, and R, different from both $\hat{P}$ and $\hat{Q}$, a point such that $\mathscr{T}^k \circ R$ is defined for all integers k and approaches, for $k \to +\infty$, one periodic orbit $\{\hat{Q}, \mathscr{T}\circ\hat{Q}, \dots, \mathscr{T}^m \circ \hat{Q} \equiv \hat{Q}\}$ and for $k \to -\infty$, another $\{\hat{P}, \mathscr{T}\circ\hat{P}, \dots, \mathscr{T}^n \circ \hat{P} \equiv \hat{P}\}$. If these two orbits coincide, R is called a homoclinic point, if they do not agree R is said to be an heteroclinic point. In case $\hat{P}, \hat{Q}$ are hyperbolic, the heteroclinic points belonging to $\hat{P}, \hat{Q}$ lie on the intersection $\{W^-_{\hat{P}} \cap W^+_{\hat{Q}} - (\hat{P}, \hat{Q})\}$ of the curve $W^-_{\hat{P}}$ of points escaping $\hat{P}$ and the curve $W^+_{\hat{Q}}$ of points approaching $\hat{Q}$ under iteration of $\mathscr{T}$.

If $W^-_{\hat{P}}$ and $W^+_{\hat{P}}$ intersect at some point $R \neq \hat{P}$, (as it was the case for the example last discussed), then R as well as all the points $\mathscr{T}^k \circ R$ are homoclinic.

The early development of the "stochastic band" in a measure-preserving diffeomorphism $\mathscr{T}$ on $\mathbb{R}^2$ is in summary characterized by an island-chain of elliptic points alternating with hyperbolic points (to which the band is associated) in a typical pattern. The stochastic scatter of return points occurs in the proximity of hyperbolic points and is due to the phenomenon of homoclinic oscillations.

The geometric structure associated with each "island" in the chain is illustrated in Fig. 14 where h_i, $i = 1,2,\dots,N$ are the hyperbolic points (fixed under some N-th iterate of $\mathscr{T}$), that alternate with an equal number of elliptic points e_i, $i = 1,\dots,N$.

We have earlier already discussed the existence, invariance and generically transversal intersection of the stable and unstable manifolds where

$$W^{\pm}_i = \left\{ x \in \mathbb{R}^2 \mid \lim_{m \to \infty} \mathscr{T}^{\pm m} x = h_i \right\} \quad . \tag{6.49}$$

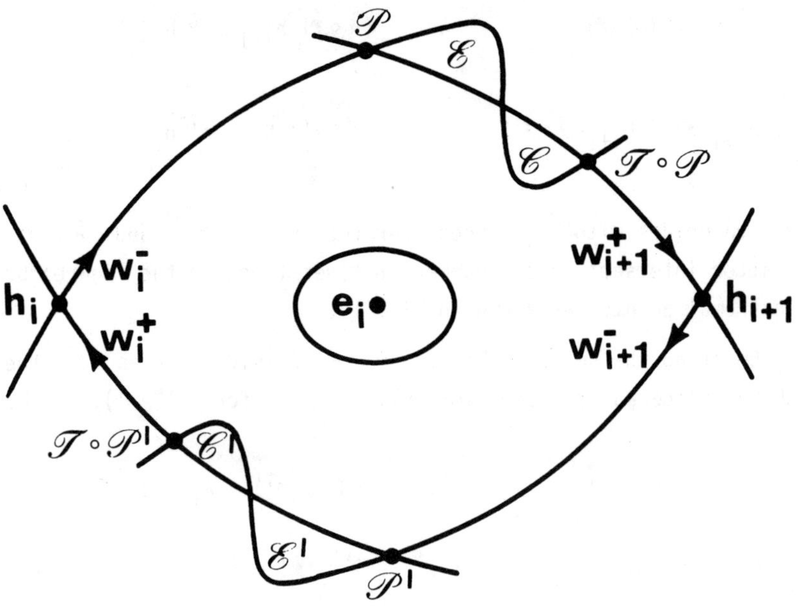

Fig. 14

If $\mathscr{P}$ is a homoclinic point in $W_i^- \cap W_{i+1}^+$, and the arcs $\widehat{h_i \mathscr{P}} \subset W_i^-$ and $\widehat{\mathscr{P} h_{i+1}} \subset W_{i+1}^+$ intersect only at $\mathscr{P}$ (and similarly for the point $\mathscr{P}' \in W_i^+ \cap W_{i+1}^-$), the island A_i becomes the domain in $\mathbb{R}^2$ with boundary

$$\sigma_i = \widehat{h_i \mathscr{P}} \cup \widehat{\mathscr{P} h_{i+1}} \cup \widehat{h_{i+1} \mathscr{P}'} \cup \widehat{\mathscr{P}' h_i} \quad . \tag{6.50}$$

One notices that $\mathscr{T} \circ A_i = \{x \in \mathbb{R}^2 | \mathscr{T}^{-1} x \in A_i\}$ is bounded by

$$\mathscr{T} \circ \sigma_i = \widehat{h_i \mathscr{T} \circ \mathscr{P}} \cup \widehat{(\mathscr{T} \circ \mathscr{P}) h_{i+1}} \cup \widehat{h_{i+1} (\mathscr{T} \circ \mathscr{P}')} \cup \widehat{(\mathscr{T} \circ \mathscr{P}') h_i} \tag{6.51}$$

which is in itself a simple closed curve. Moreover with

$$\widehat{h_i \mathscr{P}} \subset \widehat{h_i (\mathscr{T}\circ\mathscr{P})} \quad , \quad \widehat{(\mathscr{T}\circ\mathscr{P})\, h_{i+1}} \subset \widehat{\mathscr{P}\, h_{i+1}} \quad ,$$

$$\widehat{h_{i+1} \mathscr{P}'} \subset \widehat{h_{i+1} (\mathscr{T}\circ\mathscr{P}')} \quad , \quad \widehat{(\mathscr{T}\circ\mathscr{P}')\, h_i} \subset \widehat{\mathscr{P}'\, h_i} \quad , \tag{6.52}$$

$\mathscr{T}\circ\sigma_i$ coincides with σ_i except for the arcs $\widehat{\mathscr{P}\mathscr{T}\circ\mathscr{P}}$ and $\widehat{\mathscr{P}'\mathscr{T}\circ\mathscr{P}'}$. The latter intersect an odd number of times (once in the figure) between the extremal points, as shown in Fig. 14.

Denoting the set of lobes on $\widehat{\mathscr{P}\mathscr{T}\circ\mathscr{P}}$ outside A_i as $\mathscr{E}$, the set of those inside as $\mathscr{C}$ (and similarly $\mathscr{E}'$, $\mathscr{C}'$ for $\widehat{\mathscr{P}'\mathscr{T}\circ\mathscr{P}'}$), we have

$$\mathscr{E} \cup \mathscr{E}' \subset \mathscr{T}\circ A_i \quad , \quad (\mathscr{E} \cup \mathscr{E}') \cap A_i = \emptyset$$

$$(\mathscr{C} \cup \mathscr{C}') \cap \mathscr{T}\circ A_i = \emptyset \quad , \quad \mathscr{C} \cup \mathscr{C}' \subset A_i \quad . \tag{6.53}$$

Notice that the description (6.53) is characteristic of such lobes, namely no other area elements can be described in that way. Thus a set of positive measure escapes from the island at each iteration to form $\mathscr{E} \cup \mathscr{E}'$, whereas a set of equal measure $\mathscr{C} \cup \mathscr{C}'$ enters it: points can be added or escape in no other way. Then all orbits which are candidates to exhibiting truly stochastic properties such as ergodicity, mixing (and hyperbolicity) can be expected to lie within the invariant set

$$S = \bigcup_{m=-\infty}^{+\infty} \mathscr{T}^m \circ \bigcup_{i=1}^{N} \Big(\mathscr{C}(i) \cup \mathscr{C}'(i) \Big) \tag{6.54}$$

where $\mathscr{C}(i)$, $\mathscr{C}'(i)$ are the lobe sets $\mathscr{C}$ associated with the i-th island.

7. The Non-Existence of Integrals of Motion near Homoclinic Orbits and Statistical Behavior

Let us consider the following one-dimensional piece-wise linear transformation:

$$\mathscr{T}x = (-)^{[\![2x]\!]}\, 2x + \quad 1 - (-)^{[\![2x]\!]} \quad [\![2x]\!] \qquad , \tag{7.1}$$

defined for $0 \le x \le 1$. Any function $f(x)$ is changed by $\mathscr{T}$ to

$$\mathbb{T}f(x) \doteq f(\mathscr{T}^{-1} x) = \frac{1}{2}\left[f\left(\frac{x}{2}\right) + f\left(1 - \frac{x}{2}\right)\right] \qquad . \tag{7.2}$$

The two transformations $\mathbb{T}$ and $\mathscr{T}$ are then clearly related by the convolution property:

$$\int_0^1 f(x)\, g\,(\mathscr{T}x)\, dx = \int_0^1 g(x)\, \mathbb{T}\, f(x)\, dx \tag{7.3}$$

holding for any arbitrary pair of functions f and g. We utilize here $\mathbb{T}$ and $\mathscr{T}$ to define intuitively a few concepts that will be given in a more general formal way in the next chapter. $\mathscr{T}$, as defined by (7.1), is usually referred to as the baker's transformation. The pictorial idea is that to make bread a baker first prepares a paste of flour, salt, water, etc., then to make sure that the paste is homogeneous, he repeats a set of movements which may be schematized, in a one-dimensional model, as follows: he stretches the paste (whose finite length we may assume to be 1) to twice its length, then he folds it over to its original length.

This is described just by (7.1), which may be more explicitly written as

$$\mathscr{T}x = \begin{cases} 2x & \text{if} \quad 0 \le x \le \frac{1}{2} \\ 2(1-x) & \text{if} \quad \frac{1}{2} \le x \le 1 \end{cases} \qquad . \tag{7.4}$$

Functions such as $f(x)$ describe, for example, the distribution of salt in the paste. Equation (7.4) shows immediately that $x = 0$ and

$x = 2/3$ are fixed points $(\mathcal{T}x = x)$ of $\mathcal{T}$, namely they form a cycle of length one (one step).

Moreover, any rational point $x = p/q$, where p and q are integers, p being odd and $q = 2^{\alpha} \cdot q'$ (with α an integer and q' an odd integer), eventually forms a cycle of length, r, where $r \le 1/2\,(q' - 1)$. Indeed, by (7.1)

$$\mathcal{T}\frac{p}{q} = \frac{p}{2^{\alpha-1}\,q'} - \left[\!\left[\frac{p}{2^{\alpha-1}\,q'}\right]\!\right] \doteq \frac{p_1}{q/2} \quad , \tag{7.5}$$

where $p_1 \le p$ ($p_1 = p$ if $p < q/2$, $p_1 = p - q/2$ if $p > q/2$), and hence

$$\mathcal{T}^{\alpha}\ \frac{p}{q} = \frac{p'}{q'} \quad , \tag{7.6}$$

with $p' \le q'$ (as well as $p' \le p$). Further applications of $\mathcal{T}$, give

$$\mathcal{T}^{n}\ \frac{p}{q} = \frac{p^{(n)}}{q'} \quad , \quad n > \alpha \quad , \tag{7.7}$$

where $p^{(n)}$ is an even integer $< q'$. Now it is either $p^{(n)} = p^{(n-1)}$, where one hits a fixed point at the (n-1)-st step ($p^{(n-1)} = 0$ or $p^{(n-1)} = 2\,q'/3$, if q' is a multiple of 3), or $p^{(n)} \ne p^{(n-1)}$. The latter situation, with the number of non-zero even integers $< q'$ being equal to $1/2\,(q' - 1)$, can occur at most $1/2\,(q' - 1)$ times before a value of $p^{(k)}$ is re-encountered, thus giving rise to a cycle. In the baker's language, if the distribution of salt were initially concentrated at a rational point, then even after an infinite number of repetitions of $\mathcal{T}$, the salt would be found at a finite number of points.

Thus the natural question to ask is whether the operation $\mathcal{T}$ is a good transformation if one wants to have a asymptotically uniform distribution. This can be formulated by imposing the requirement,

$$\lim_{n\to\infty} \mathbb{T}^n f(x) = \text{const.} = \int_0^1 f(x)\, dx \quad . \tag{7.8}$$

We will now show that if $f(x)$ has support of a non-zero measure (i.e. $f(x) \neq 0$ over an interval however small of $0 \vdash\!\!\dashv 1$, which excludes concentration at rational points) then (7.8) is always satisfied. In fact, write $f(x)$ first as a Fourier series

$$f(x) = \sum_{m=-\infty}^{+\infty} f_m \exp(i2\pi mx) \tag{7.9}$$

where

$$f_m = \int_0^1 f(x) \exp(-\, i2\pi mx)\, dx \quad . \tag{7.10}$$

Notice then that, by (7.2)

$$\mathbb{T} \cos(\pi mx) = \begin{cases} \cos\left(\frac{1}{2}\pi mx\right) & \text{if} \quad m \text{ is even} \\ 0 & \text{if} \quad m \text{ is odd} \end{cases} \tag{7.11}$$

and moreover

$$\cos(\pi mx) = \mathbb{T} \exp(i2\pi mx) \tag{7.11'}$$

so that the right-hand side of (7.11) represents $\mathbb{T}^2 \exp(i2\pi mx)$. Thus, if $m = 2^{\alpha}m'$ (with α, m' integers and m' odd), $\mathbb{T}^{\alpha+2} \exp(i2\pi mx) = 0$. Now finally,

$$\begin{aligned} \lim_{n\to\infty} \mathbb{T}^n f(x) &= \sum_{m=-\infty}^{+\infty} f_m \lim_{n\to\infty} \mathbb{T}^n \exp(i2\pi mx) \\ &= f_0 = \int_0^1 f(x)\, dx \end{aligned} \tag{7.12}$$

and this proves (7.8). By (7.3) the above property (usually referred to as complete mixing) can be written in the form

$$\lim_{n\to\infty} \int_0^1 f(x)\, g(\mathcal{T}^n x)dx = \int_0^1 f(x)dx \int_0^1 g(x)dx \tag{7.13}$$

for any pair of functions f and g. In several applications one is interested in a less restrictive requirement than (7.13) simply called mixing (not complete mixing). The latter asks that the mapping, say $\bar{\mathcal{T}}$, be such that

$$\lim_{n\to\infty} \int_0^1 x^m \bar{\mathcal{T}}^n x\, dx = [2(m+1)]^{-1} \tag{7.14}$$

for every positive integer m. For analytic functions $f(x)$, (7.14) corresponds to (7.13) with $g(x) = x$. Since, by Carlson's theorem, (7.14) remains true for any real positive m, only the functions $f(x)$ not regular enough to be expanded as infinite sums of real powers of x are excluded. Thus one may expect that mixing and complete mixing are connected properties.

For the baker's transformation $\mathcal{T}$ of (7.1), it is easy to verify the mixing property. In fact this is somewhat superfluous since the complete mixing property implies mixing, yet it is extremely instructive. One subdivides first the interval $0 \vdash\!\dashv 1$ into 2^n equal subintervals of width 2^{-n}, denoting by $x_k = k \cdot 2^{-n}$, $k = 0,1,\dots,2^n$, their end-points. Any integral over the unit interval can now be decomposed as follows:

$$\int_0^1 f(x)dx = \sum_{k=0}^{2^n-1} \int_{x_k}^{x_{k+1}} f(x)dx =$$

$$= \sum_{k=0}^{2^{n-1}-1} \left(\int_{x_{2k}}^{x_{2k+1}} f(x)dx + \int_{x_{2k+1}}^{x_{2k+2}} f(x)dx \right) . \tag{7.15}$$

In the application we are considering, $f(x) \doteq x^m \mathcal{T}^n x$, and it is straightforward to check by iteration of (7.4), that

$$\mathcal{T}^n x = \begin{cases} 2^n(x - x_{2k}) & \text{for} \quad x_{2k} \leq x \leq x_{2k+1} \\ 2^n(x_{2k+2} - x) & \text{for} \quad x_{2k+1} \leq x \leq x_{2k+2} \end{cases} ,$$

$$k = 0,1,\ldots,2^{n-1} - 1 \quad . \tag{7.16}$$

Thus one can finally write (7.14) (for $\bar{\mathcal{T}} = \mathcal{T}$) in the form

$$\int_0^1 x^m \mathcal{T}^n x \, dx = 2^n \sum_{k=0}^{2^{n-1}-1} \left(\int_{x_{2k}}^{x_{2k+1}} x^m (x - x_{2k}) dx + \right.$$

$$\left. + \int_{x_{2k+1}}^{x_{2k+2}} x^m (x_{2k+2} - x) dx \right) =$$

$$= 2^{-n(m+1)} \int_0^1 d\xi \cdot \xi \sum_{k=0}^{2^{n-1}-1} \left[(2k + \xi)^m + (2k + 2 - \xi)^m \right] \tag{7.17}$$

where an obvious change of variables has been performed. The integral in (7.17) is easily computed and gives

$$\int_0^1 x^m \mathcal{T}^n x \, dx = \frac{2^{-n(m+1)+1}}{(m+1)(m+2)} \left[2^{m+3} \sum_{k=1}^{2^{n-1}-1} k^{m+2} - \sum_{k=1}^{2^n-1} k^{m+2} \right] +$$

$$+ \frac{2^n}{(m+1)(m+2)} \quad . \tag{7.18}$$

On using the known formula,

$$\sum_{k=1}^{N-1} k^p = \frac{1}{p+1} \sum_{r=1}^{p} \binom{p+1}{r} B_r \, N^{p+1-r} \tag{7.19}$$

where $\{B_r\}$ are the Bernoulli's numbers, (7.18) rewrites

$$\int_0^1 x^m \mathscr{T}^n x \, dx = \frac{1}{m+1} \sum_{r=0}^{m} \binom{m+1}{r} \frac{B_{r+2}}{(r+1)(r+2)} \, 2^{-nr+1} (2^{r+2} - 1) \quad . \tag{7.20}$$

A simple algebra shows then at last that for a large n the right-hand side of (7.20) has asymptotic form $\sim \frac{1}{2(m+1)} + O(2^{-2n})$, proving, as expected, mixing according to the criterion of (7.14). Let now A denote a finite or denumerable set of symbols (or integers) that will be referred to as an alphabet. Consider the doubly infinite sequences

$$s \doteq \{\ldots s_{-2}, s_{-1}, s_0 \quad ; \quad s_1, s_2, \ldots\} \tag{7.21}$$

of elements $s_k \in A$ as points in a space, S. S can be given a topology for instance by defining as the neighbourhood basis of a given point $s^* \in S$, the sets

$$N_j \doteq \{s \in S | \; s_k = s_k^* \quad ; \quad |k| < j\} \quad , \quad j = 1,2,3,\ldots \tag{7.22}$$

Moreover S can be made a measure space in the following way: assign to every element $a \in A$ a real-positive number p_a, $0 \leq p_a \leq 1$, such that

$$\sum_{a \in A} p_a = 1 \tag{7.23}$$

and define, for the sets

$$E_{\{k_i\ ;\ i=1,\dots,r\}}\left(\{a_i\ ;\ i=1,\dots,r\}\right) \doteq \left\{s\ \ S\ ;\ s_{k_i} = a_i\ | i=1,\dots,r\right\} \quad , \tag{7.24}$$

the Borel measure

$$\mu\left(E_{\{k_i\}}(\{a_i\})\right) = \prod_{i=1}^{r} p_{a_i} \quad . \tag{7.25}$$

S, together with the Borel algebra generated by the sets $E_{\{k\}}(\{a\})$ and the measure μ, constitutes a measure space. On the measure space S as defined, the shift homeomorphism σ is

$$(\sigma(s))_k = s_{k-1} \quad , \quad s \in S \quad . \tag{7.26}$$

The mapping σ obviously preserves the measure μ: it is called the Bernoulli shift. On considering the special case of a binary alphabet, $A \doteq \{0\,,1\}$ and selecting $p_0 = p_1 = 1/2$, one can associate the resulting mapping σ with a two-dimensional baker's transformation. Indeed, associate with any $s \in S$, the two real positive numbers, $x\,,y$, both less than 1, represented by the binary forms

$$x \doteq \sum_{k=-\infty}^{0} s_k\, 2^{k-1} \quad , \quad y = \sum_{k=1}^{\infty} s_k\, 2^{-k} \quad . \tag{7.27}$$

Equation (7.27) obviously maps a point $s \in S$ onto a point of the square of side 1 on the $(x\,,y)$-plane, $Q = \{x\,,y\ |0 \le x \le 1\ ;\ 0 \le y \le 1\}$: denoted by τ such a mapping. Clearly τ takes μ into the Lebesgue measure in Q, so that the mapping induced by σ in Q, namely $\mathcal{T} = \tau\sigma\tau^{-1}$ is a measure preserving transformation in Q. Upon defining

$$\mathcal{T}'\begin{pmatrix} x \\ y \end{pmatrix} = \begin{pmatrix} x' \\ y' \end{pmatrix} \quad , \tag{7.28}$$

one finds

$$x' = 2x - s_0 \quad , \quad y' = \frac{1}{2}(y + s_0) \quad . \tag{7.29}$$

However,

$$s_0 = 2x - \sum_{k=1}^{\infty} s_{-k}\, 2^{-k} = [\![2x]\!] \tag{7.30}$$

and (7.29) can be written

$$x' = 2x - [\![2x]\!] \quad , \quad y' = \frac{1}{2}\left(y + [\![2x]\!]\right) \quad . \tag{7.31}$$

The latter generalizes to two dimensions, the baker's transformation (7.1) (to which the first of (7.31) is isomorphic). Indeed it has the following geometric interpretation. Consider the two halves of Q,

$$Q_1 \doteq \left\{x\,,y \mid 0 \le x \le \frac{1}{2} \;,\; 0 \le y \le 1\right\} \quad ,$$

$$Q_2 \doteq \left\{x\,,y \mid \frac{1}{2} \le x \le 1 \;,\; 0 \le y \le 1\right\} \quad . \tag{7.32}$$

These are mapped by $\mathscr{T}'$ respectively onto

$$\mathscr{T}' \circ Q_1 = \left\{x\,,y \mid 0 \le x \le 1 \quad , \quad 0 \le y \le \frac{1}{2}\right\} \quad ;$$

$$\mathscr{T}' \circ Q_2 = \left\{x\,,y \mid 0 \le x \le 1 \quad , \quad \frac{1}{2} \le y \le 1\right\} \quad . \tag{7.33}$$

In other words globally $\mathscr{T}' \circ Q_1 \cup \mathscr{T}' \circ Q_2 = Q = Q_1 \cup Q_2$, but the rectangles Q_1 and Q_2 have been first stretched by a factor 2 in the horizontal direction (and compressed by a factor 1/2 in the vertical direction, since the Lebesgue measure is conserved) and then piled up again, as shown in Fig. 15. Since the baker's transformation has been shown to be complete-mixing, so is the Bernoulli shift σ; which was just seen to be isomorphic to it.

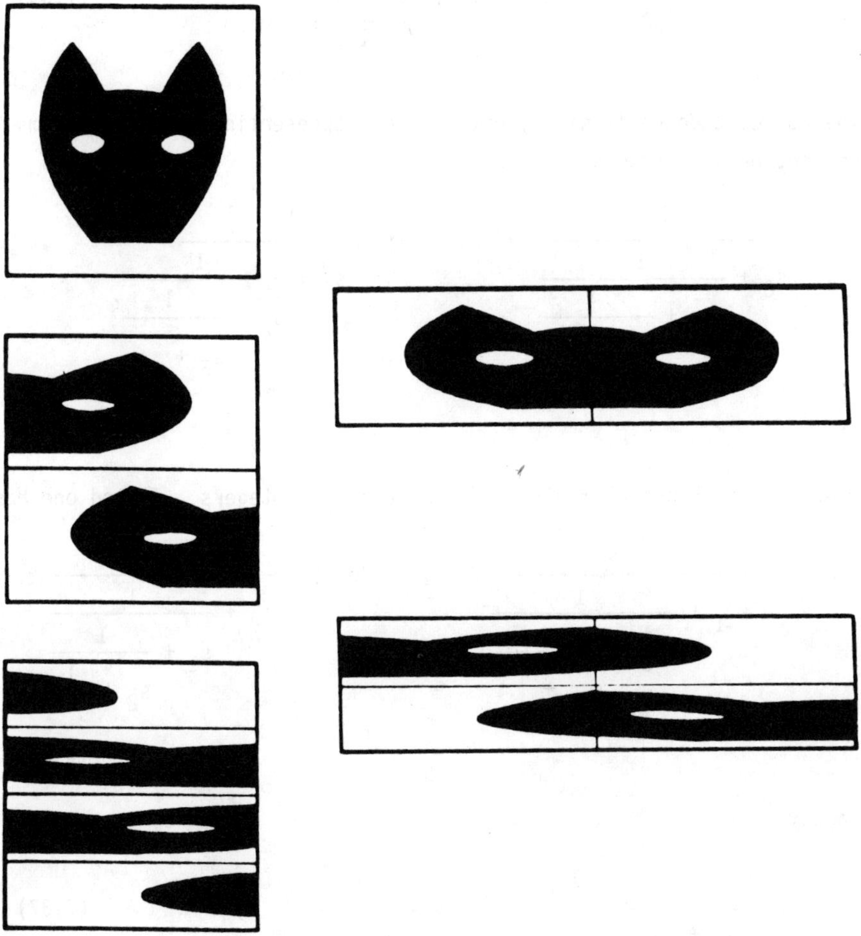

Fig. 15.

Another mapping $\mathscr{T}'' : Q \to Q$ which is related to the shift σ in S is the following:

$$x' = \frac{1}{x} - \left[\!\left[\frac{1}{x} \right]\!\right] \quad , \qquad y' = \frac{1}{y + [\![\frac{1}{x}]\!]} \quad . \tag{7.34}$$

The correspondence is simply obtained by representing x, y in terms of continued fractions

$$x = \cfrac{1}{s_0 + \cfrac{1}{s_{-1} + \cfrac{1}{s_{-2} + \cfrac{1}{\cdots}}}} \quad , \qquad y = \cfrac{1}{s_1 + \cfrac{1}{s_2 + \cfrac{1}{s_3 + \cdots}}} \quad , \tag{7.35}$$

where the alphabet elements s_k are positive integers. Indeed one has

$$\sigma \circ x = \cfrac{1}{s_{-1} + \cfrac{1}{s_{-2} + \cfrac{1}{s_{-3} + \cdots}}} \quad , \qquad \sigma \circ y = \cfrac{1}{s_0 + \cfrac{1}{s_1 + \cfrac{1}{s_2 + \cfrac{1}{\cdots}}}} \quad , \tag{7.36}$$

namely,

$$x = \frac{1}{s_0 + \sigma \circ x} \quad , \qquad \sigma \circ y = \frac{1}{s_0 + y} \quad , \tag{7.37}$$

or

$$\sigma \circ x = \frac{1}{x} - s_0 \quad , \qquad \sigma \circ y = \frac{1}{y + s_0} \quad . \tag{7.38}$$

On the other hand, from (7.35)

$$s_0 = \frac{1}{x} - \cfrac{1}{s_{-1} + \cfrac{1}{s_{-2} + \cfrac{1}{\cdots}}} = \left[\!\left[\frac{1}{x} \right]\!\right] \tag{7.39}$$

and (7.38) are thus equivalent to (7.34), provided

$$x' = \sigma \circ x \quad , \quad y' = \sigma \circ y \quad . \tag{7.40}$$

Notice however that σ as defined by (7.38) is not a Bernoulli shift; the measure it preserves is not a product measure as required by (7.25) and besides, it is a single valued but not one-to-one mapping. In fact, by writing $\mu(x\,,y) = \rho(x\,,y)dxdy$, where ρ is the measure density, one checks right away that

$$\rho(x\,,y) = \frac{1}{\ln 2(1 + xy)^2} \tag{7.41}$$

gives the invariant measure under σ: indeed, from (7.40), (7.38)

$$\left| \frac{dx'}{dx} \right| = \frac{1}{x^2} \quad , \quad \left| \frac{dy'}{dy} \right| = \frac{1}{(y + s_0)^2} \tag{7.42}$$

and

$$\frac{dx'dy'}{(1 + x'y')^2} = \frac{dxdy}{x^2(y + s_0^2)\left[1 + (\frac{1}{x} - s_0)(\frac{1}{y + s_0})\right]^2} =$$

$$= \frac{dxdy}{(1 + xy)^2} \quad . \tag{7.43}$$

The coefficient in (7.41) guarantees that

$$\int_Q \rho(x, y)dxdy = 1 \qquad . \tag{7.44}$$

On the other hand, consider the first map in (7.34), $\mathscr{T}''_x : x \to \frac{1}{x} - \left[\!\left[\frac{1}{x} \right]\!\right]$. Write then $\left[\!\left[\frac{1}{x} \right]\!\right] = s_0 \doteq n$, and denote the corresponding value of x by x_n, $\frac{1}{n+1} \leq x_n < \frac{1}{n}$. The inverse image of a given point $x_0 \in [0, 1]$ is given by

$$\frac{1}{x_n} - n = x_0 \tag{7.45}$$

i.e.

$$x_n = \frac{1}{x_0 + n} \qquad . \tag{7.46}$$

Since n can be any positive integer, $n \geq 1$, this implies that there are infinitely many (countable) values of x whose image is x_0. Notice also that $\mathscr{T}''_{x_n} \circ \left[\frac{1}{n+1}, \frac{1}{n}\right] = [0, 1]$, i.e. on any interval $\left[\frac{1}{n+1}, \frac{1}{n}\right]$ there is precisely one point mapped to x_0.

There is a relation between the ergodic properties of maps such as (7.31) or (7.34) and those of geodesic flows on the two-dimensional surfaces of constant negative curvature. Consider indeed the hyperbolic plane $|\mathbb{H}| = \{x + iy \mid y > 0\}$. The metric on $|\mathbb{H}|$ is given by $ds^2 = \frac{dx^2 + dy^2}{y^2}$, and the geodesics for it are the half-circles and straight lines orthogonal to the x-axis. Let $\mathbb{U}$ be the unit tangent bundle consisting of unit tangent vectors on $|\mathbb{H}|$. $\mathbb{U}$ can be coordinatized by $\vec{u} \equiv (x, y, \theta)$, where (x, y) is the base point of $\vec{u} \in \mathbb{U}$ and θ is the angle measured counter-clockwise between the positive x-axis and $\vec{u}$.

A geodesic flow is defined as the class of homeomorphisms of $\mathbb{U}$

$$\Phi_t : \vec{u} \to \vec{u}_t \quad , \quad -\infty < t < \infty \tag{7.47}$$

where $\vec{u}, \vec{u}_t$ are respectively the initial and terminal unit tangent vectors to a geodesic segment of signed length t. Φ_t has a particularly simple description if one resorts to the following coordinates. For each $\vec{u} \in \mathbb{U}$, let γ be the geodesic determined in $|\mathbb{H}|$. Then assign to $\vec{u}$ the coordinates of the two intersection points of γ with the x-axis, say η, ξ (where ξ denotes e.g. the point in the forward direction), as well as the hyperbolic distance s measured along γ from the initial point. In these coordinates one can write:

$$\Phi_t : (\eta, \xi, s) \to (\eta, \xi, s+t) \quad . \tag{7.48}$$

The hyperbolic measure associated with the geodesic flow (invariant under Φ_t) is simply

$$d\mu = \frac{d\eta \, d\xi \, ds}{(\eta - \xi)^2} \quad . \tag{7.49}$$

Let Γ be the modular group, i.e. the group of transformations of $|\mathbb{H}|$ into itself, defined by

$$\Gamma : z \to \tau(z) \quad , \quad z = x + iy \in |\mathbb{H}| \tag{7.50}$$

where

$$\tau(z) = \frac{az + b}{cz + d} \quad , \quad ad - bc = 1 \quad , \quad a, b, c, d \in \mathbb{Z} \quad . \tag{7.51}$$

Γ acts on $\mathbb{U}$ as well, the action on the latter being

$$\Gamma : (x, y, \theta) \to \left(\mathrm{Re}\ \tau(z), \mathrm{Im}\ \tau(z), \theta + \arg(\frac{d\tau}{dz})\right) \quad . \tag{7.52}$$

For simplicity we refer to Γ as G when acting on $\mathbb{U}$. The fundamental domain for Γ is the hyperbolic triangle

$$\mathscr{F} = \left\{ z \in |\mathbb{H}| \;\middle|\; |z| > 1 \quad , \quad -\frac{1}{2} < x < \frac{1}{2} \right\} \tag{7.53}$$

with $\bar{\mathscr{F}}$ as its closure ($>$ and $<$ in (7.53) replaced by $\geq , \leq$ respectively). $|\mathbb{H}|$ is tesselated with the images of $\mathscr{F}$ under Γ,

i) for any two distinct elements $\tau_1 , \tau_2 \in \Gamma$,

$$\tau_1(\mathscr{F}) \cap \tau_2(\mathscr{F}) = \emptyset \tag{7.54}$$

ii) $$\bigcup_{\tau \in \Gamma} \tau(\bar{\mathscr{F}}) = |\mathbb{H}| \quad . \tag{7.55}$$

Observe moreover that the two elementary transformations

$$A : z \to \alpha(z) \quad , \quad \alpha(z) = z - 1$$

$$B : z \to \beta(z) \quad , \quad \beta(z) = -\frac{1}{z} \tag{7.56}$$

generate the whole Γ. In particular, any two points $\xi , \eta \in \mathbb{R}$ are mapped one into the other by the transformation

$$g_\gamma = A^{n_1} B A^{-n_2} B \ldots A^{(-)^{p+1} n_p} A^{(-)^q m_q} B \ldots A^{m_2} B A^{-m_1} \tag{7.57}$$

where the sets of integers $\{n_i ; i = 1,\ldots,p\}$, $\{m_j ; j = 1,\ldots,q\}$ are defined by the two sequences uniquely expressing ξ and η respectively in the form of continued fractions:

$$\xi = n_1 + \cfrac{1}{n_2 + \cfrac{1}{n_3 + \cfrac{1}{n_4 + \ldots}}} \quad ,$$

$$\eta = m_1 + \cfrac{1}{m_2 + \cfrac{1}{m_3 + \cfrac{1}{m_4 + \cdots}}} \quad . \tag{7.58}$$

[Notice that rational numbers correspond to terminating sequences; real numbers to infinite sequences.] The integers p and q in (7.57) are defined by the requirement that $n_{p+r} = m_{q+r}$, $\forall r \geq 0$. In order to prove that $\xi = g_\gamma \eta$, one only needs to derive from (7.58) — recalling that, by (7.56), $A^n : z \to z - n$, $B^{-1} = B$ — the following relations:

$$A^{-n_1} \circ \xi = \cfrac{1}{n_2 + \cfrac{1}{n_3 + \cfrac{1}{n_4 + \cdots}}} \quad ,$$

$$BA^{-n_1} \circ \xi = - \left(n_2 + \cfrac{1}{n_3 + \cfrac{1}{n_4 + \cdots}} \right) \quad ,$$

$$A^{n_2}BA^{-n_1} \circ \xi = - \cfrac{1}{n_3 + \cfrac{1}{n_4 + \cdots}} \quad ,$$

$$BA^{n_2}BA^{-n_1} \circ \xi = n_3 + \cfrac{1}{n_4 + \cdots} \quad , \ldots \tag{7.59}$$

together with the analogous expressions for η (the m_α's replacing the n_α's). If the "tails" of ξ and η agree — corresponding to the condition $n_{p+r} = m_{q+r}$, $\forall r \geq 0$ — (7.59)'s imply (7.57). g_γ is manifestly of the form (7.50), (7.51) with

$$a = x_p\, z_{q-1} + x_{p-1}\, z_q \quad ,$$

$$b = x_p\, w_{q-1} + x_{p-1}\, w_q \quad ,$$

$$c = y_p\, z_{q-1} + y_{p-1}\, z_q \quad ,$$

$$d = y_p\, w_{q-1} + y_{p-1}\, w_q \quad , \qquad (7.60)$$

where $\{x_k\,, y_k\,, z_k\,, w_k\ ;\ k \geq 1\}$ are the solutions of the recursion relations:

$$x_{k+1} = (-)^k\, n_{k+1}\, x_k - x_{k-1} \quad , \quad x_0 = 1 \quad , \quad x_1 = n_1$$

$$y_{k+1} = (-)^k\, n_{k+1}\, y_k - y_{k-1} \quad , \quad y_0 = 0 \quad , \quad y_1 = -1$$

$$z_{n+1} = (-)^{k+1}\, m_{k+1}\, z_k - z_{k-1} \quad , \quad z_0 = 0 \quad , \quad z_1 = -1$$

$$w_{k+1} = (-)^{k+1}\, m_{k+1}\, w_k - w_{k-1} \quad , \quad w_0 = 1 \quad , \quad w_1 = -m_1 \quad .$$

(7.61)

Moreover the opposite vertical boundary lines of $\mathscr{F}$ (see (Fig. 16) are mapped into each other by α, whereas the bottom boundary is globally fixed under β (though the unit circle arc that it corresponds to, is reflected with respect to the Imz-axis). We define the space $|\mathbb{H}|$ modulo such two transformations a modular surface, and denote it as M (in other words M can be thought of as $\bar{\mathscr{F}}$ with the boundaries identified in the above sense).

In a perfectly analogous way, unit vectors with base points on the boundaries of $\mathscr{F}$ can be identified under the action of α and β, thought of as transformations of G. We denote by $\mathscr{M}$ the space $\mathbb{U}$ modulo such identification. M and $\mathscr{M}$ can be coordinated by introducing the projection maps,

$$\pi : |\mathbb{H}| \to M \quad , \quad \pi(z) = \Gamma \circ z \quad , \quad z \in |\mathbb{H}|$$

$$p : \mathbb{U} \to \mathscr{M} \quad , \quad p(\vec{u}) = G \circ \vec{u} \quad , \quad \vec{u} \in \mathbb{U} \qquad (7.62)$$

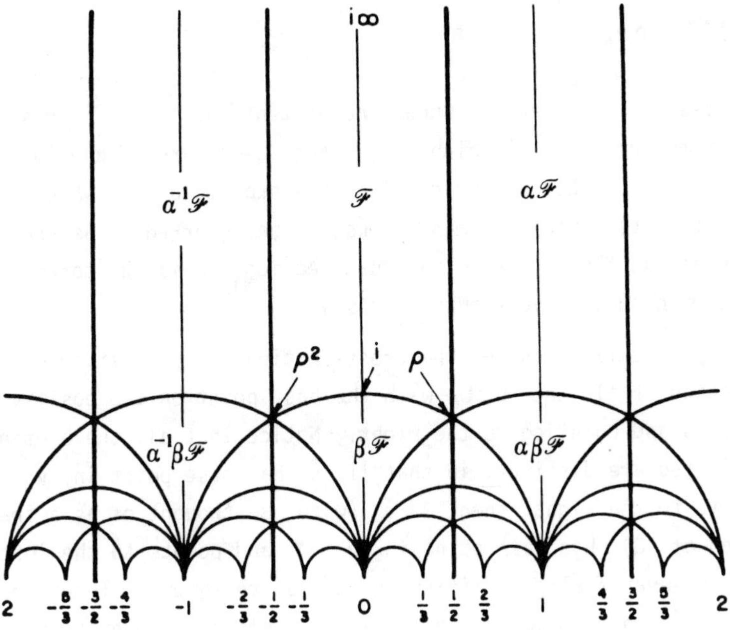

$\varrho = \exp(i\pi/3)$

Fig. 16.

where the right-hand side denote Γ- and G-orbits of z and $\vec{u}$ respectively. π and p are locally one-to-one, with the exception of the points i and $\frac{1+i\sqrt{3}}{2}$ for the Γ-orbit, and of the unit vectors based at the same points for the G-orbit. Thus (x, y) [or (η, ε)] provide local coordinates at the points of $\mathscr{M}\backslash\left\{\pi(i), \pi(\frac{1+i\sqrt{3}}{2})\right\}$ and (x, y, θ) [or (η, ξ, s)] provide local coordinates at the points of $\mathscr{M}\backslash\left\{p(0, 1, \theta), p(\frac{1}{2}, \frac{\sqrt{3}}{2}, \theta), 0 \leq \theta < 2\pi\right\}$. Avoiding the exceptional points the formula for $d\mu$ then carries over to $\mathscr{M}$. Since Φ_t commutes with G on $\mathbb{U}$, $\mathscr{M}$ inherits the geodesic flow

$$\Phi_t^{(\mathscr{M})} = p\,\Phi_t\, p^{-1} \tag{7.63}$$

and because μ is invariant under the elements of G, $\Phi_t^{(\mathscr{M})}$ has the invariant measure $\mu^{(\mathscr{M})}$, which is but the μ-measure of any local inverse under p. Let us now consider the subsets of $\mathscr{M}$ which every $\Phi_t^{(\mathscr{M})}$-orbit meets infinitely many times, to be referred to as cross-sections of the flow on $\mathscr{M}$. The cross-section map is the correspondence between successive return points.

In particular consider the cross-section S consisting of the p-projection of all the $\vec{u} \in \mathbb{U}$ with the base point on the positive y-half-axis and pointing to the right. Notice that all the elements thus selected are distinct, in that if $\vec{u}$ has base point in $y_+ \cap \beta\circ\mathscr{F}$, and points to the right, then $\beta\circ\vec{u}$ (with β thought of as a transformation of G) has base point in $y_+ \cap \mathscr{F}$ and points to the left. There is however a slight difficulty one has to remove. There are indeed $\Phi_t^{(\mathscr{M})}$-orbits which do not visit S infinitely many times, namely the p-projections of orbits starting (or terminating) at cusp points. The latter are described by $p(\eta, \xi, s)$, $-\infty < s < \infty$, with either η or ξ rational (or ∞). The difficulty can be removed by restricting one's attention to the case when all η and ξ's under consideration are irrational; such a choice does not affect the generality of the results one obtains since the cases thus discarded encompass a set of measure zero.

Provided the above restrictions are kept in mind, S can be simply described by assigning to each $\vec{v} = p\circ\vec{u} \in S$ its η, ξ coordinates:

$$S \equiv \left\{(\eta, \xi) \mid \xi > 0 \quad , \quad \eta < 0\right\} \quad . \tag{7.64}$$

S can be decomposed as:

$$S = S_1 \cup S_2 \quad ,$$

$$S_1 \equiv \left\{(\eta, \xi) \mid 0 < \xi < 1 \quad , \quad \eta < 0\right\} \quad ,$$

$$S_2 \equiv \left\{(\eta, \xi) \mid \xi > 1 \quad , \quad \eta < 0\right\} \quad . \tag{7.65}$$

To obtain an expression for the cross-section map $\mathscr{T}_s$, consider the hyperbolic triangle Δ with vertices $0, 1, \infty$. The three sides of Δ are equivalent under Γ, the sides $\widehat{01}, \widehat{1\infty}$ being carried respectively into $\widehat{0\infty}$ by the transformations $\alpha \circ \beta \circ \alpha : z \to \dfrac{z}{1-z}$ and $\alpha : z \to z - 1$. Moreover, Γ-orbits of internal points of Δ are certainly distinct from those of the boundary points.

Let γ again be the geodesic determined by $\vec{u} \in \mathbb{U}$.

i) If $\vec{v} \in S_2$, then γ leaves Δ through the right vertical wall. Let (η, ξ, s) be the unit tangent at the point of departure. γ is identified under α with a geodesic entering Δ at the left vertical wall with the unit tangent at the point of entrance being $(\eta - 1, \xi - 1, s)$. Hence

$$\mathscr{T}_s(\eta, \xi) = (\eta - 1, \xi - 1) \quad , \quad \text{on } S_2 \quad . \tag{7.66}$$

ii) If $\vec{v} \in S_1$, then γ leaves Δ through the base $\widehat{01}$ with unit tangent (η, ξ, s). γ is identified under $\alpha \circ \beta \circ \alpha$ with a geodesic entering Δ at the left vertical wall with unit tangent $\left(\dfrac{\eta}{1-\eta}, \dfrac{\xi}{1-\xi}, s\right)$. Hence

$$\mathscr{T}_s(\eta, \xi) = \left(\frac{\eta}{1-\eta}, \frac{\xi}{1-\xi}\right) \quad , \quad \text{on } S_1 \quad . \tag{7.67}$$

It appears from (7.66) and (7.67) that the invariant measure $d\mu_s$ for $\mathscr{T}_s$ can be obtained simply from $d\mu$ as given in (7.49) but by dropping ds

$$d\mu_s = \frac{d\eta d\xi}{(\eta - \xi)^2} \quad , \tag{7.68}$$

realizing that the map

$$\mathscr{T} \circ \xi = \begin{cases} \dfrac{\xi}{1-\xi} & 0 < \xi < 1 \\ \xi - 1 & 1 < \xi < \infty \end{cases} , \tag{7.69}$$

is a factor map of $\mathscr{T}_S$. The invariant measure $d\nu$ for $\mathscr{T}$, can be written in the form

$$d\nu = h(\xi)d\xi \tag{7.70}$$

where $h(\xi)$ is obtained by integration of $d\mu_s$ with respect to η:

$$h(\xi) = \int_{-\infty}^{0} \frac{d\eta}{(\eta - \xi)^2} = \frac{1}{\xi}, \quad \xi > 0 \quad . \tag{7.71}$$

Let finally $n = n(\xi)$ be (for $0 < \xi < 1$) the smallest positive integer such that the n-th iterate $\mathscr{T}^n$ is in $[0, 1]$, and consider the map f:

$$f(\xi) \doteq \mathscr{T}^n \circ \xi \quad , \quad 0 < \xi < 1 \quad . \tag{7.72}$$

It is straightforward to check that

$$f(\xi) = \frac{\xi}{1-\xi} - [\![\frac{\xi}{1-\xi}]\!] = \frac{1}{1-\xi} - [\![\frac{1}{1-\xi}]\!] , \quad 0 < \xi < 1 \tag{7.73}$$

with measure $\frac{d\xi}{\xi}$, by (7.70), (7.61). Upon setting $\xi = \frac{1}{2}(1-x)$, the mapping induced by (7.73), $\tilde{f} : x \to \frac{1}{x} - [\![\frac{1}{x}]\!]$ is seen to coincide with $\mathscr{T}_x''$, and to have indeed the density of the invariant measure given by $\rho(x) = \frac{1}{1+x}$.

An alternative of the latter formula (which is due to Gauss), can be obtained by writing (resorting to (7.46))

$$\sum_{n=1}^{\infty} \frac{dx_n}{1+x_n} = \sum_{n=1}^{\infty} \frac{dx_n}{1+\dfrac{1}{x_0+n}} = \sum_{n=1}^{\infty} \frac{x_0+n}{x_0+n+1} \cdot x_n^2\, dx_0 =$$

$$= \sum_{n=1}^{\infty} \frac{1}{(x_0+n+1)(x_0+n)}\, dx_0 =$$

$$= \sum_{n=1}^{\infty} \left(\frac{1}{n+x_0} - \frac{1}{n+x_0+1} \right) dx_0 = \frac{dx_0}{1+x_0} \tag{7.74}$$

which proves the invariance of the density $\rho(x) = \frac{1}{1+x}$ (based on the property stated above that there is an infinite number of images x_n for each x_0). This completes the proof that the map $\mathscr{T}_x''$ is indeed a factor map of a cross-section map for the geodesic flow on the (unit tangent bundle of the) modular surface.

8. Stability of Dynamical Systems

The concepts mentioned so far can be discussed in a global framework by studying the dynamical systems in a context somewhat autonomous i.e. focusing one's attention on the flow connected with them rather than on the details of their dynamics.

Let then M be a smooth manifold, and $\vec{X}$ a vector field on it (no assumption need to be made either on the dimensions of M or on its compactness). A dynamical system ϕ on M is the flow defined by the differential equations

$$\dot{\vec{x}} = \vec{X} \quad , \quad \vec{x} \in M \quad . \tag{8.1}$$

In other words ϕ is a smooth map

$$\phi : \mathbb{R} \times M \to M \tag{8.2}$$

such that for all $t \in \mathbb{R}$,

$$\phi_t : M \to M \tag{8.3}$$

is a diffeomorphism where

$$\phi_t(\vec{x}) = \phi(t, \vec{x}) \tag{8.4}$$

and therefore

$$\phi_s \circ \phi_t = \phi_{s+t} \quad . \tag{8.5}$$

Consider now the set $\mathscr{X}$ of all vector fields on M and equip it with a topology and an equivalence relation as follows. The topology is C^1, i.e. two vector fields $\vec{X}, \vec{X}' \in \mathscr{X}$ are close if both the vectors and their (partial) derivatives are close. $\vec{X}, \vec{X}'$ are defined as equivalent if there exists a homeomorphism of M sending globally the orbits of $\vec{X}$ onto the orbits of $\vec{X}'$. These two properties bring us to the concept of structural stability: the dynamical system ϕ is said to be <u>structurally stable</u> if the corresponding vector field $\vec{X}$ has a neighbourhood in $\mathscr{X}$ all the elements of which are its equivalents. Structural stability is different from Lyapunov stability. The latter refers to individual orbits and requires that in a given dynamical system ϕ orbits corresponding to slightly perturbed initial conditions are close in M. The former refers to the whole dynamical system and asks that if ϕ is perturbed into a ϕ' close to ϕ in $\mathscr{X}$, the global quality of the entire flow (i.e. of all the orbits, allowing for arbitrary initial conditions) is preserved.

Point $\vec{x} \in M$ is said to be wandering if there exist a submanifold N of M including $\vec{x}$ and a time t_0 such that

$$N \cap \phi_t(N) = \emptyset \quad , \quad \forall\, t > t_0 \quad . \tag{8.6}$$

Denote by Ω the set of non-wandering points: Ω contains the attractors i.e. the points and sets toward which all orbits eventually flow. Its connected components are the basins of attraction for the various attractors. For gradient systems, for which

$$\vec{X} = -\operatorname{grad} V \tag{8.7}$$

where $V : M \to \mathbb{R}$ is a scalar "potential" function, Ω is a finite set of fixed points (extrema or saddles of V) and closed orbits.

As an example, let us start from the usual harmonic oscillator (for simplicity we assume $n = 1/2 \dim M = 1$), which we write — in suitably rescaled variables — in the following way. Let the manifold M (in this case non-compact) be $\mathbb{C}$ parametrized in terms of $z, \bar{z}$ (where we set $z = x + i\dot{x}$, $x \in \mathbb{R}$). Then $\vec{x} = \binom{z}{\bar{z}}$ is a complex 2-vector and if we assume

$$V = \frac{i}{2}(z^2 - \bar{z}^2) \tag{8.8}$$

then (8,7) gives

$$\vec{X} = i \begin{pmatrix} -z \\ \bar{z} \end{pmatrix} \tag{8.9}$$

whereby (8.1) reads $\dot{z} = -iz$ ($\dot{\bar{z}} = i\bar{z}$), namely

$$\ddot{x} + x = 0 \quad . \tag{8.10}$$

The solution of (8.10) corresponding to initial conditions $x(0) = x_0$, $\dot{x}(0) = \dot{x}_0$, is

$$x = x_0 \cos t + \dot{x}_0 \sin t \tag{8.11}$$

or

$$z = (x_0 + i\dot{x}_0)\, e^{-it} \quad . \tag{8.12}$$

The orbit of this flow in $\mathbb{C}$ are circles concentric with the origin

$$z\bar{z} = |z|^2 \equiv x^2 + \dot{x}^2 = x_0^2 + \dot{x}_0^2 = \rho_0^2 = \text{const.} \tag{8.13}$$

Upon identifying M with $\mathbb{R}^2$ (with coordinates $x, \dot{x}$), the flow induced by (8.10) can be seen to be Lyapunov stable because small changes in $x_0, \dot{x}_0$ amount to a small change in the orbit radius ρ_0. However it is structurally unstable with respect to the small structural perturbation $(0 < \varepsilon << 1)$ of the form

$$\vec{X} \to \vec{X}_\varepsilon = \vec{X} + \varepsilon \begin{pmatrix} -z + \bar{z} \\ z - \bar{z} \end{pmatrix} \tag{8.14}$$

or equivalently

$$V \to V_\varepsilon = V + \frac{1}{2}\varepsilon(z - \bar{z})^2 \tag{8.15}$$

corresponding to a small damping. Indeed in such a case, (8.10) is replaced by

$$\ddot{x} + 2\varepsilon\dot{x} + x = 0 \tag{8.16}$$

whose solution, to the order ε, is

$$x = e^{-\varepsilon t}[x_0 \cos t + (\dot{x}_0 + \varepsilon x_0)\sin t] \quad . \tag{8.17}$$

The orbits are then spirals, asymptotically approaching the origin, no matter how small ε is. They are of different global structure from the orbits of (8.10).

It is interesting that the harmonic oscillator is structurally stable under more complicated perturbations. For example the

Van der Pol oscillator, corresponding to the perturbed form of Eq. (8.10):

$$\ddot{x} + \varepsilon(x^2 - 1)\dot{x} + x = 0 \quad , \tag{8.18}$$

where again $0 < \varepsilon << 1$, has a non-wandering set for the resulting flow on $\mathbb{R}^2$ which consists of a single point (the origin which is a repeller point) and an attracting limit cycle, close to a circle of radius $\sqrt{2}$. This can be easily checked by referring $\mathbb{R}^2$ to polar coordinates (ρ, θ):

$$\begin{aligned} x &= \rho \cos \theta \\ \dot{x} &= \rho \sin \theta \end{aligned} \tag{8.19}$$

and rewriting then (8.18) in the form

$$\rho \cos \theta(\dot{\theta} + 1) + \sin \theta[\dot{\rho} + \varepsilon\rho(\rho^2 \cos^2 \theta - 1)] + \\ + \varepsilon\rho^2\dot{\rho} \cos^3 \theta = 0 \quad . \tag{8.20}$$

Equation (8.20) suggests a solution of the form

$$\theta = \theta_0 - t + 0(\varepsilon^2) \tag{8.21}$$

(i.e. $\dot{\theta} + 1 = 0$), where

$$\theta_0 = \text{arc tan} \frac{\dot{x}_0}{x_0} \quad . \tag{8.22}$$

ρ is then a solution of

$$\left(1 + \varepsilon\rho^2 F(\theta_0 - t)\right) \dot{\rho} + \varepsilon\rho \left(\rho^2 \cos^2(\theta_0 - t) - 1\right) = 0 \tag{8.23}$$

with initial condition $\rho(0) = \rho_0$, where $F(x) = \dfrac{\cos^3 x}{\sin x}$. Noticing that, for $\varepsilon = 0$, (8.23) becomes $\dot{\rho} = 0$, one looks now for a solution in the form of a power series in ε. Up to the first order in ε, namely setting

$$\rho = \rho_0 + \varepsilon\rho_1 + O(\varepsilon^2) \tag{8.24}$$

(8.23) reads — to the desired order —

$$\dot{\rho}_1 = -\rho_0[\rho_0^2 \cos^2(t - \theta_0) - 1] \tag{8.25}$$

whose solution (with $\rho_1(0) = 0$) is

$$\rho_1(t) = \rho_0 \left(1 - \frac{1}{2}\rho_0^2\right)t - \frac{1}{4}\rho_0^3 \sin\,[2(t - \theta_0)] \tag{8.26}$$

which proves the statement.

Let us now consider the parametrized Van der Pol equation:

$$\ddot{x} + \varepsilon(x^2 - b)\dot{x} + x = 0 \quad . \tag{8.27}$$

Repeating the whole argument one can easily check that when $b < 0$ the flow in the phase plane $\mathbb{R}^2$ has only one attractor point at the origin. When $b > 0$ the origin turns into a repeller and an attracting limit cycle of radius $\sqrt{2b}$ appears. In other words the non-wandering set in the new phase space $\mathbb{R}^2 \times \mathbb{R}$ consists of a paraboloid and its axis. The change in global behaviour happening at $b = 0$ is referred to as Hopf bifurcation.

Next, let us consider what happens when the damping becomes large (we replace ε by κ, to remind that it is now $>>1$), and rescale once more the variables so that (8.27) writes

$$\ddot{x} + \kappa(3x^2 - b)\dot{x} + x = 0 \quad . \tag{8.28}$$

It is convenient to use in such a case, a different phase space, with

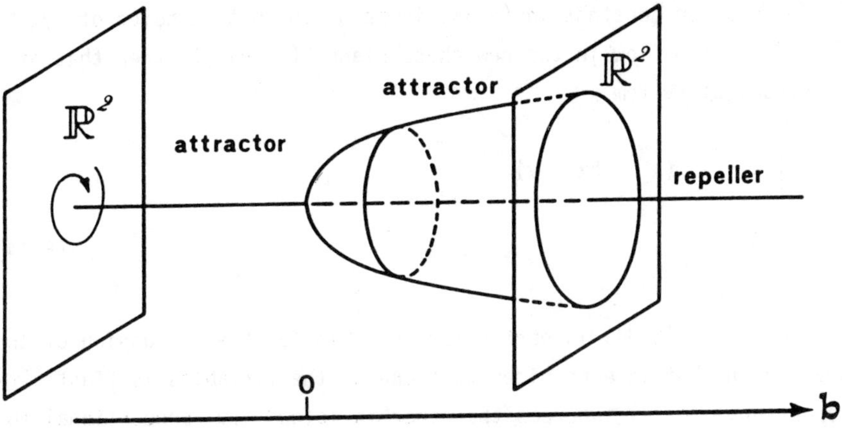

Fig. 17.

coordinates x and $\int^t x\,dt$, in that $\dot{x}$ is no longer a good geometrical way to represent the oscillator (it may become very large). Let us then introduce the variable

$$y(t) = y_0 - \frac{1}{\kappa}\int_0^t x(t)dt \tag{8.29}$$

where

$$y_0 = y(0) = x_0^3 - bx_0 - \frac{1}{\kappa}\dot{x}_0 \quad . \tag{8.30}$$

Clearly

$$\dot{y} = -\frac{1}{\kappa}x \tag{8.31}$$

and the oscillator Equation (8.28) becomes

$$\ddot{x} + \kappa(3x^2\dot{x} - b\dot{x} - \dot{y}) = 0 \tag{8.32}$$

which, integrated once with respect to time, gives

$$\dot{x} + \kappa(x^3 - bx - y) = \text{const.} = 0 \quad . \tag{8.33}$$

The integration constant in (8.33) is zero due to the choice of y_0 in (8.30). The flow in the new phase plane (x, y) is then that of the dynamical system

$$\begin{cases} \dot{x} = -\kappa(x^3 - bx - y) \\ \dot{y} = -\frac{1}{\kappa}x \end{cases} \qquad . \tag{8.34}$$

The form (8.34) is particular suitable for the discussion of the dynamics in that it emphasizes that one of the variables is "fast" ($\dot{x}$ proportional to κ) whereas the other is "slow" ($\dot{y}$ proportional to κ^{-1}). Consider now the curve in $\mathbb{R}^2$ (referred to as <u>slow manifold</u>)

$$y = x^3 - bx \tag{8.35}$$

and allow b to vary as well (i.e. consider again the extended phase space $\mathbb{R}^2 \times \mathbb{R}$ of Fig. 17). Equation (8.35) represents a surface Γ in the (x, y, b)-space which is but the canonical cusp catastrophe surface in the sense of R. Thom. The slow manifolds are the intersections of the surface Γ with planes $b = \text{constant}$. Off Γ, due to the first equation in (8.34) the orbits are nearly parallel to the x-axis (up to $O(1/\kappa)$). This implies that the upper and lower sheets of Γ $(3x^2 > b)$ act as attractors, whereas the middle sheet $(3x^2 < b)$ is a repeller [see Fig. 18].

When the point gets close to an attracting sheet of Γ, the orbit tends to remain along such sheet as long as it does not cross one of the fold curves $3x^2 = b$. Should that happen, then the orbit has an abrupt catastrophic jump onto the other sheet, where once more the point flows slowly along the slow manifold. If $b < 0$, the origin $x = y = 0$ is a unique attractor point. When $b > 0$, the origin turns into a repeller and the system exhibits an hysteresis cycle, consisting of two portions of slow flow along the upper and lower attracting sheets of Γ, alternating with two fast catastrophic jumps between the two sheets.

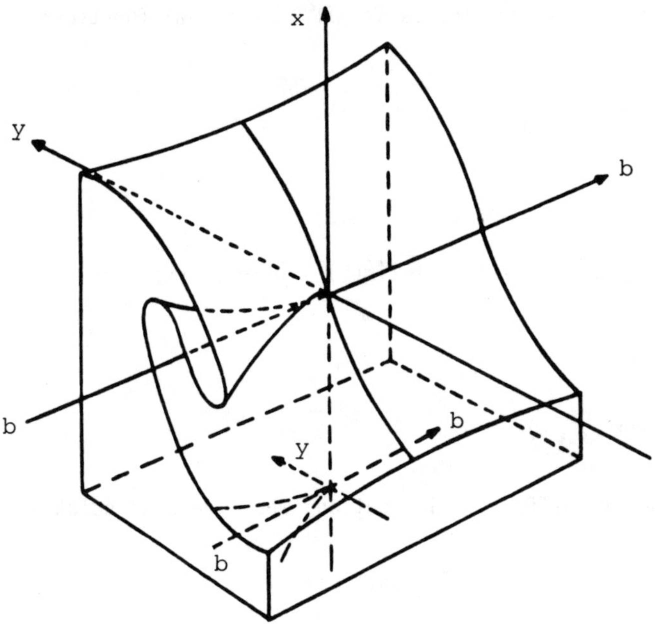

Fig. 18.

It is interesting to notice that if for any dynamical system one can record the (periodic) behaviour along a closed equilibrium orbit, then it is always possible to write a second order differential equation having such an orbit as (unique) attractor. Indeed, let C be the orbit, in some manifold M. By suitably scaling the time unit, one can always assume that C has a period 2π. Let T denote the circle $\mathbb{R} (\mathrm{mod}\ 2\pi)$, representing periodic time, and $T \to C$ be the diffeomorphism giving timing along the orbit. Finally let $M \to \mathbb{R}$ be the given measurement, and denote by ζ the (periodic) state variable under consideration

$$\zeta : T \to C \subseteq M \to \mathbb{R}$$

$$\zeta = \zeta(t) \quad , \quad t \in T \quad , \quad \zeta \in \mathbb{R} \quad . \tag{8.36}$$

Assuming $x, \dot{x}$ as coordinates in $\mathbb{R}^2$, the real function ψ is defined,

$$\psi : \mathbb{R}^2 \times T \to \mathbb{R} \tag{8.37}$$

given by

$$\psi = \psi(x, \dot{x}, t) = \ddot{\zeta}(t) + 2[\dot{\zeta}(t) - \dot{x}] + 2[\zeta(t) - x] \quad . \tag{8.38}$$

Then

$$\ddot{x} = \psi(x, \dot{x}, t) \tag{8.39}$$

is the required differential equation, whose general solution is

$$x = x(t) = \zeta(t) + Ae^{-t} \cos(t - \phi) \tag{8.40}$$

where A and ϕ are real constants, decays indeed — as desired — to $\zeta(t)$.

Returning to the analysis of the non-wandering set, it remains to be mentioned that the connected components of Ω (referred to as basic sets of the dynamical system) may not be manifold. The most fundamental example of such possibility, due to S. Smale, is the so-called horseshoe, which occurs as a saddle in a three-dimensional flow (connected e.g. to perturbed versions of the forced Van der Pol equation), and turns out to have just the structure we associated, in previous section, with the stochastic features of mixing and ergodicity.

What Smale did was first to define a diffeomorphism Φ mapping a subset of the plane — say the unit square $Q = \{(x, y) \in \mathbb{R}^2 \mid |x| \leq 1/2 , |y| \leq 1/2\}$ — into the plane $\mathbb{R}^2$ in the following way.

i) Φ maps Q into the region bounded by dotted lines in Fig. 19 [such that $\Phi(P_i) = P_i'$, $i = 1,2,3,4$] diffeomorphically.

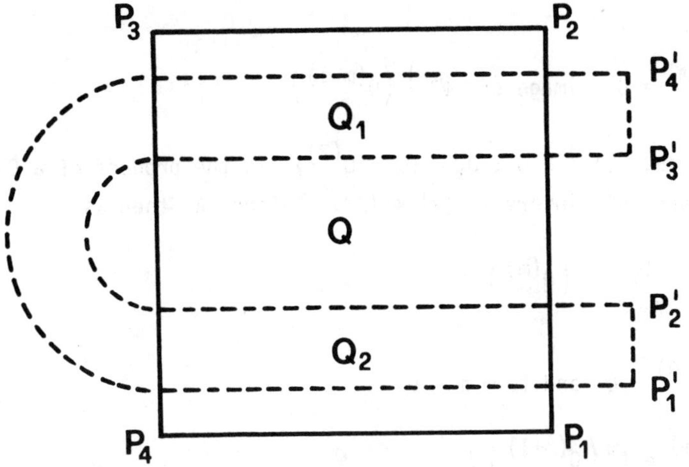

Fig. 19.

ii) On each component R_1, R_2 of $\Phi^{-1}(\Phi(Q) \cap Q)$, Φ is a linear map, i.e. $R_i = \Phi^{-1}(Q_i)$, $i = 1,2$ as seen in Fig. 20.

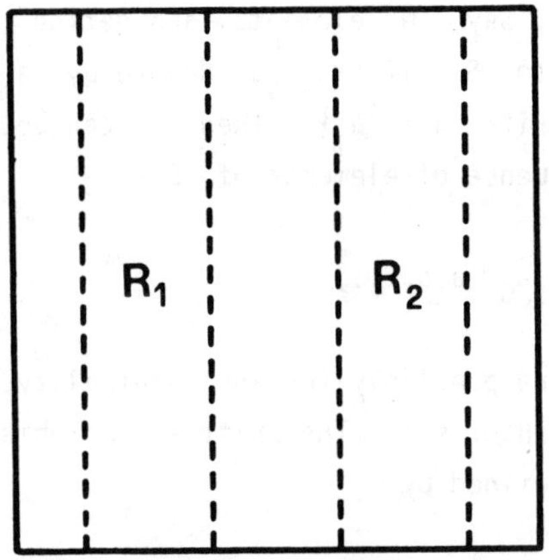

Fig. 20.

Then the intersections of all the images

$$\bigcap_{m=1}^{\infty} \Phi^m \left(Q^{(m)} \right) \cap Q \quad , \tag{8.41}$$

where

$$Q^{(m)} = Q \cap \text{image of } \Phi^{m-1}\left(Q^{(m-1)}\right) \tag{8.42}$$

(in Fig. 19, $Q^{(1)} \equiv Q$; $Q_1 \cup Q_2 = Q^{(2)}$) is the product of a Cantor set and the unit interval $|x| \leq 1/2$. Define Λ then as

$$\Lambda = \bigcap_{m \in \mathbb{Z}} \Phi^m\left(Q^{(m)}\right) \tag{8.43}$$

where $Q^{(0)} \equiv Q$ and for $m < 0$

$$Q^{(m)} = \Phi^m\left(Q^{(m+1)}\right) \quad . \tag{8.44}$$

Λ can be thought of as the set of non-wandering points of $\Phi : Q \to \mathbb{R}^2$. Clearly $\Lambda (\subseteq \Omega)$ is compact and invariant under Φ.

Successively Smale proved that Φ is topologically conjugate with the shift atomorphism. The latter is defined as follows. Let S be a finite set (of, say, N elements) and define τ_s as the set of functions from $\mathbb{Z}$ to S. If $a \in \tau_s$, denote by $a_m \in s$ the value of a at $m \in Z$, and write $a = \{a_m\}$. Then a can be thought of as a doubly infinite sequence of elements of S:

$$a = (\ldots a_{-2} a_{-1} a_0 \mid a_1 a_2 \ldots) \quad . \tag{8.45}$$

For general S (more precisely for any cardinality N of S), τ_s is homeomorphic to a Cantor set. The shift automorphism σ of τ_s is a map $\sigma : \tau_s \to \tau_s$ defined by

$$[\sigma(a)]_m = a_{m+1} \quad , \tag{8.46}$$

which shifts the bar at the right-hand side of (8.45), one position to the right.

The topological equivalence of Φ and σ is far reaching. Indeed one can derive from it that for any perturbation Ψ of Φ Λ'

— defined similarly — is also compact and invariant under Ψ, $\Psi : \Lambda' \to \Lambda'$ is conjugate to the shift σ. This implies that starting from Φ, one can extend it in infinitely many ways (e.g. letting the image of $\Phi(Q)$ wind half-way or several times around Q before intersecting it a second time, as in Figs. 21 and 22; or having it pass through Q several times, as in Figs. 23 and 24) without changing $\Phi : \Lambda \to \Lambda$ in the first case, or just changing the cardinality of S (which is the number of components of $Q \cap \Phi(Q)$) in the second, in the neighbourhood of Λ (of course the global extension of Φ outside Ω will change).

The proof of the stochastic features of Φ is thus reconducted to the case $N = 2$ (i.e. the a_i can be chosen to be 0's and 1's, and a, associated to any binary digit) which was thoroughly discussed with the map (7.31).

Technically the proof requires — to be rigorously complete — that one must show that it is possible to embed the square Q in the sphere $S^{(2)}$, so that Φ may be used to define a global diffeomorphism $D : S^{(2)} \to S^{(2)}$ and one can think of Q as a (non-invariant) neighbourhood of a piece of the set of non-wandering points of D.

It is interesting to sketch at least some of the features of the latter construction, which allows Φ to be identified with the dynamics of homoclinic oscillations. Put first the square Q into a disc $\mathcal{D}^{(2)} \subset \mathbb{R}^2$, and extend Φ to $\Phi_0 : \mathcal{D}^{(2)} \to \mathcal{D}^{(2)}$ by mapping diffeomorphically (see Fig. 25) G to G' and F to F'. The map $\Phi_0 : F \to F'$ is defined in such a way that it is a contraction about a fixed point $P_0 \in F'$ (i.e. an homeomorphism whose eigenvalues have all absolute values less than 1). Thus Φ_0 will be a diffeomorphism of $\mathcal{D}^{(2)}$ onto a subset of $\mathcal{D}^{(2)}$ itself, and the non-wandering set is the disjointed union of Λ and P_0.

Finally one can easily extend Φ_0 to $D : S^{(2)} \to S^{(2)}$, the only further requirement being that there is an expanding fixed point P of D outside D (eigenvalues in modulus larger than 1).

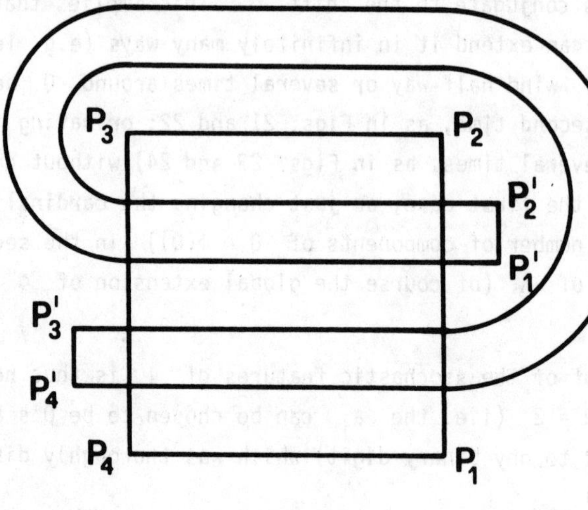

Fig. 21.

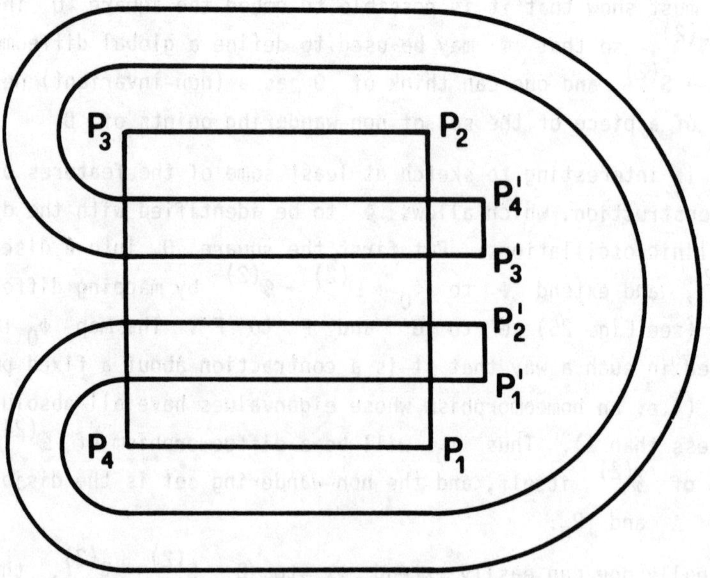

Fig. 22.

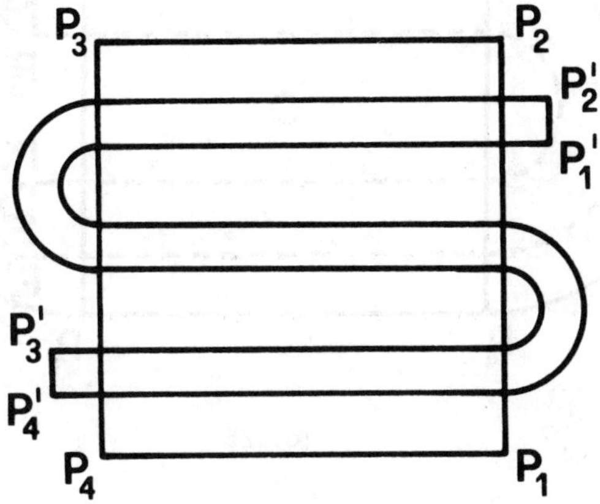

Fig. 23.

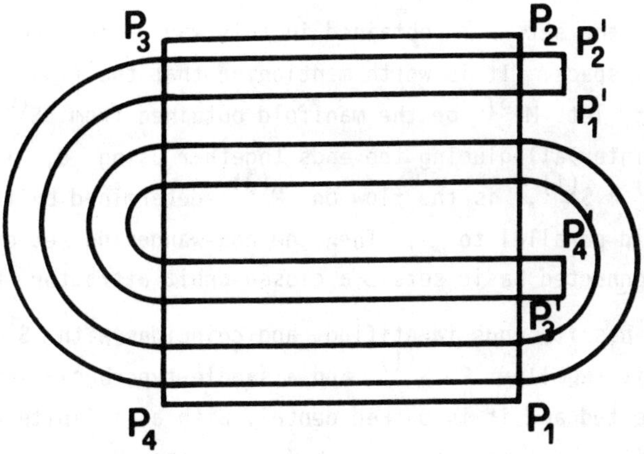

Fig. 24.

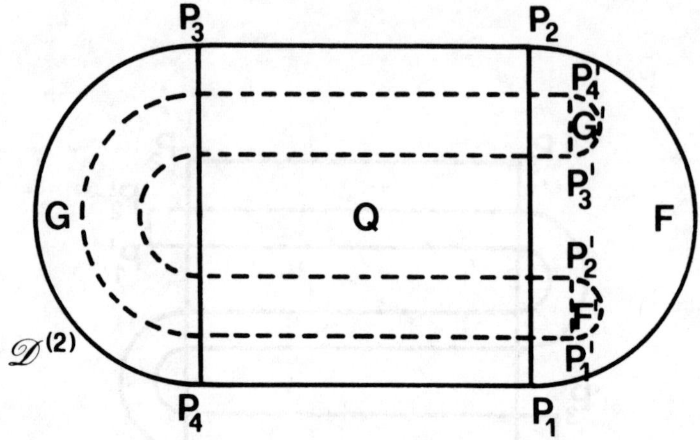

Fig. 25

The set of non-wandering points is given by

$$\Omega = \Lambda \cup P_0 \cup P \quad . \tag{8.47}$$

Notice that the set Λ — obtained in this way — is a non-connected topological space. It is worth mentioning that the flow ϕ obtained as follows: let $M^{(3)}$ be the manifold obtained from $S^{(2)} \times \mathscr{T}$ ($\mathscr{T}$ = unit interval) glueing the ends together using D, i.e. $M^{(3)} = S^{(2)} \times S^{(1)}$, is the flow on $M^{(3)}$ determined by the unit vector field parallel to $\mathscr{T}$. Then the non-wandering set of ϕ consists of three connected basic sets: a closed-orbit attractor $P_0 \times \mathscr{T}$, (where $\mathscr{T}$ has its ends identified, and coincides with $S^{(1)}$), a closed-orbit repeller $P \times S^{(1)}$ and a saddle-type basic set $\Gamma = \Lambda \times S^{(1)}$. Γ is connected and it is packed densely with an infinite number of closed orbits whose points in Λ have parameter sequences of 0's and 1's containing as subsequences all possible finite sequences.

Return now to the complicated orbit structure of homoclinic points, with its faster and faster oscillations of W^- as it approaches $\hat{P}$, as shown in Fig. 13. It is clear that what is occurring there is the same phenomenon shown in Fig. 19 and its iterations. One can think of embedding Fig. 13 into Fig. 19, and use the horseshoe dynamics to get a satisfactory picture of the orbit structure and stability.

9. Strange Attractors and Chaos

Consider first a discrete time dynamical system Φ in which the state of the system is specified by an m-component vector $\vec{x} = (x_1(t),\ldots,x_m(t))$, the evolution by a continuous vector $\vec{X} = (X_1(\vec{x}),\ldots,X_m(\vec{x}))$, and the time scale $t \in \mathbb{Z}$. Instead of (8.1), we write

$$\Phi : x_j(t+1) = X_j\left(x_1(t),\ldots,x_m(t)\right) \quad , \qquad j = 1,\ldots,m \qquad (9.1)$$

with initial conditions $x_j(0) = x_0^{(j)}$, $j = 1,\ldots,m$. The hypothesis that the functions X_j have continuous derivatives identifies (9.1) as a differentiable dynamical system.

An interesting example of (9.1) is constructed assuming $m = 2$ and selecting again (Hénon)

$$\begin{aligned} X_1 &= x_2 + 1 - ax_1^2 \\ X_2 &= bx_1 \end{aligned} \quad , \qquad (9.2)$$

(compare with (6.29) with the following identifications: $p^{(k)} = x_1(t)$, $q^{(k)} = x_2(t)$, $V(q) = \frac{1}{3}\frac{a}{b^2}q^3 - q$, $\frac{S}{m\Omega} = b$, $mS\Omega = -1$; $t \equiv k$). Equation (9.1) thus takes the form

$$\begin{aligned} x_1(t+1) &= x_2(t) + 1 - a[x_1(t)]^2 \\ x_2(t+1) &= bx_1(t) \end{aligned} \quad . \qquad (9.3)$$

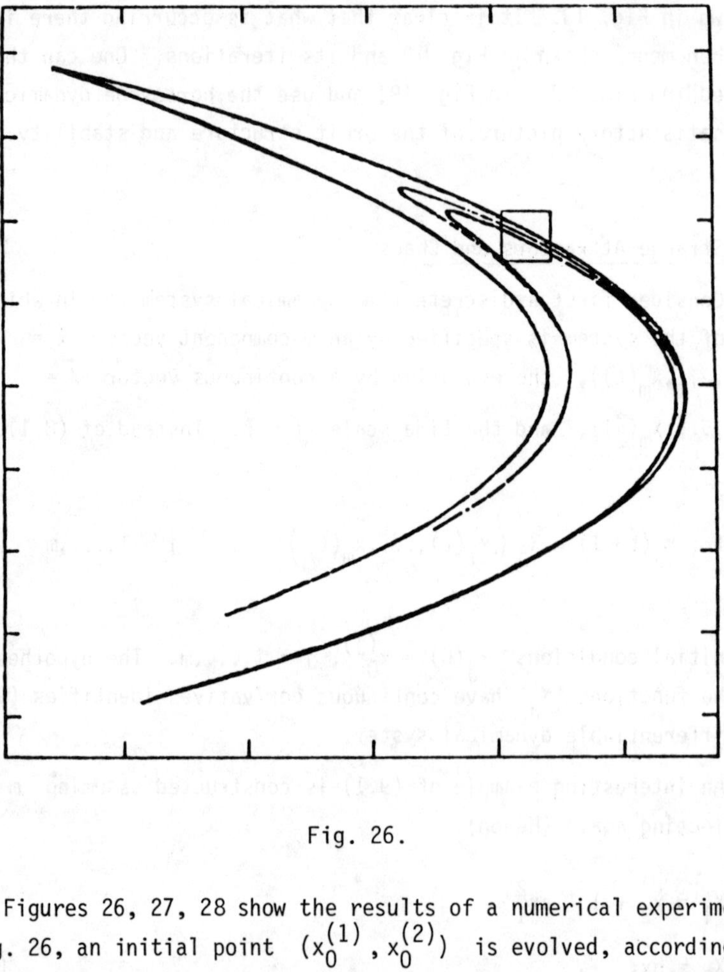

Fig. 26.

Figures 26, 27, 28 show the results of a numerical experiment. In Fig. 26, an initial point $(x_0^{(1)}, x_0^{(2)})$ is evolved, according to (9.3) for 10 000 steps with $a = 1.4$, $b = 0.3$. The figure then shows the points $x_1(t), x_2(t)$ at each step, which distribute themselves on a complex system of lines. Figure 27 exhibits the magnification of the little square in Fig. 26; a further magnification of the little square in Fig. 27 would give again a similar picture. The system of lines constitutes what is called a <u>strange attractor</u>. In Fig. 28 the same operation done in Fig. 26 is repeated, with $a = 1.3$ (b still equal to 0.3); the strange attractor disappears and is replaced by a discrete periodic attractor (seven points).

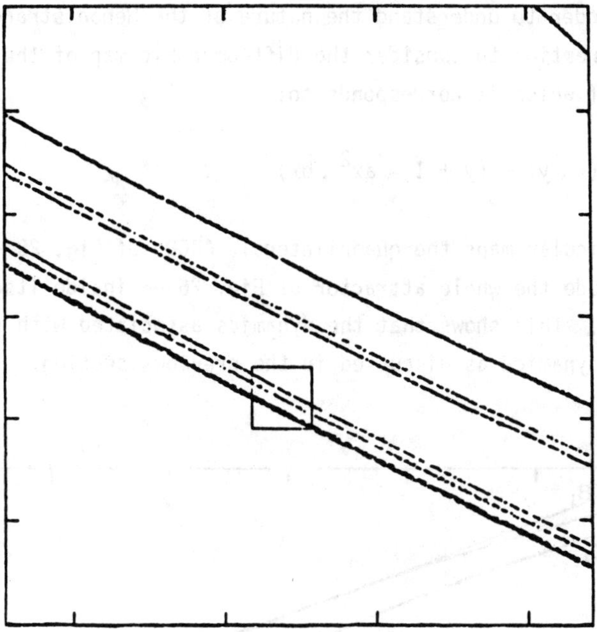

Fig. 27.

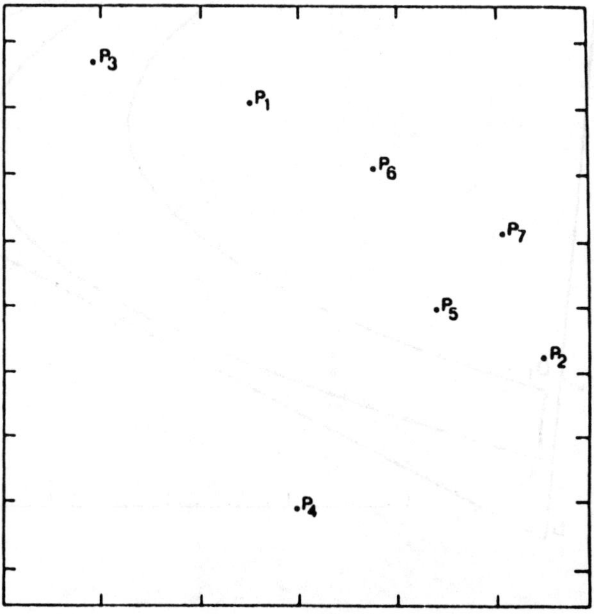

Fig. 28.

In order to understand the nature of the Hénon strange attractor, it is interesting to consider the diffeomorphic map of the plane $\mathbb{R}^2$ into itself which it corresponds to:

$$\phi : (x , y) \to (y + 1 - ax^2 , bx) \quad . \tag{9.4}$$

ϕ in particular maps the quadrilateral ABCD of Fig. 29 — chosen so as to include the whole attractor of Fig. 26 — inside itself into $A_1 B_1 C_1 D_1$. This shows that the dynamics associated with (9.3) is the horseshoe dynamics as discussed in the previous section.

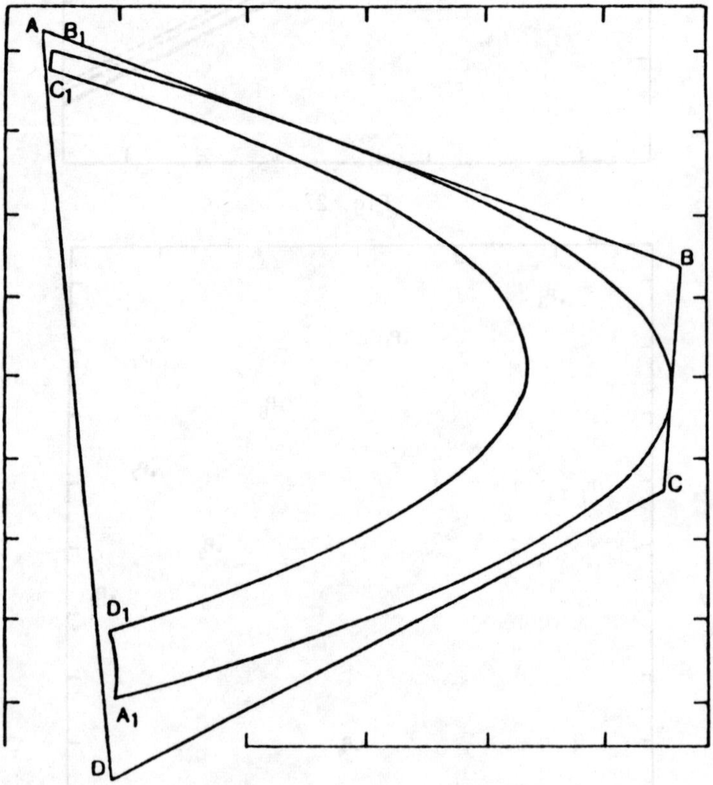

Fig. 29.

Moreover, if $P_t = (x_1(t), x_2(t))$ and $P'_t = (x'_t(t), x'_2(t))$ correspond to initial data $P_0 = (x_0^{(1)}, x_0^{(2)})$ and $P'_0 = (x_0'^{(1)}, x_0'^{(2)})$ respectively, where the initial points are close to each other, one finds from Fig. 26, and the relative numerical results that the distance $d(P_t, P'_t)$ in general increases with time (to become one of the order of the total size of the attractor) exponentially:

$$d(P_t, P'_t) \sim \alpha^t \cdot d(P_0, P'_0) \quad , \tag{9.5}$$

where $\alpha > 1$.

Thus even if the evolution described by Eqs. (9.1) is deterministic, a small error or lack of precision in the initial conditions can produce a substantially different asymptotic behaviour. $\lambda = \ln\alpha$ is referred to as a characteristic exponent, in that it characterizes the rate of such an exponential divergence with time. Equation (9.5) expresses qualitatively the high sensitivity of the dependence on initial conditions typical of the dynamics of systems exhibiting a strange attractor and thus leading to their chaotic behaviour.

We can now give a general definition of a strange attractor. With reference to Eq. (9.1), one can say that a bounded set $\mathscr{A}$ in the m-dimensional state-space is a strange attractor for the map X, if one can find a set U in the same space with the following properties:

i) U is a (m-dimensional) neighbourhood of $\mathscr{A}$; namely for each point $P \in \mathscr{A}$, there is a little ball, centered at P, entirely contained in U. Thus the whole $\mathscr{A}$ is contained in U;

ii) For every initial point $P_0 \in U$, the point P_t defined by the iteration of Eq. (9.1) remains in U for all positive t's. If t is large enough, P_t becomes and remains arbitrarily close to $\mathscr{A}$ (i.e. $\mathscr{A}$ is an attractor);

iii) If $P_0 \in U$, there is a sensitive dependence on the initial conditions, as defined by (9.5) (this makes $\mathscr{A}$ strange);

iv) $\mathscr{A}$ is indecomposable, i.e. it cannot be split into two different attractors; for each point P in $\mathscr{A}$, one can always choose a point $P_0 \in \mathscr{A}$ such that for the same positive t, P_t is arbitrarily close to P.

In the case of the Hénon dynamical system (9.3), U can be identified with the quadrilateral ABCD in Fig. 29.

Notice that the whole discussion can be extended to differentiable dynamical systems in which time is continuous $(t \in \mathbb{R})$ and (9.1) is replaced by

$$\frac{d}{dt} x_j(t) = F_j\left(x_1(t),\ldots,x_m(t)\right) \quad , \quad j = 1,\ldots,m \quad . \tag{9.6}$$

A famous example is the Lorenz system, which has $m = 3$ and

$$F_1 = -\sigma(x_1 - x_2)$$

$$F_2 = -x_1x_2 + rx_1 - x_2$$

$$F_3 = x_1x_2 - bx_3 \quad . \tag{9.7}$$

Figure 30 shows the corresponding strange attractor (with $\sigma = 10$, $r = 28$, $b = 8/3$; $P_0 \equiv (0, 0, 0)$). There is a sensitive dependence on the initial conditions.

In principle, m can be taken as infinite, and Eq. (9.6) describe then a macroscopic system such as a fluid. Moreover, the functions F at the right-hand side of (9.6) may depend on (a set of) parameters μ ranging over some domain in a "control-space" $\mathscr{M}$. Varying μ changes, in general, the structure of the attracting set, which can be constituted by a discrete set of equilibrium points, of attracting limit cycles or by an irregular set (Cantor-like). In the latter case the motion has the stochastic features referred to as <u>chaos</u>. The subsets in $\mathscr{M}$ crossing which makes the attracting set vary its nature are called <u>bifurcation</u> sets as in the case of the Hopf bifurcation studied in connection with the Van der Pol equation (Fig. 17) (there $\mu \equiv b$, $\mathscr{M} \equiv \mathbb{R}$; the bifurcation

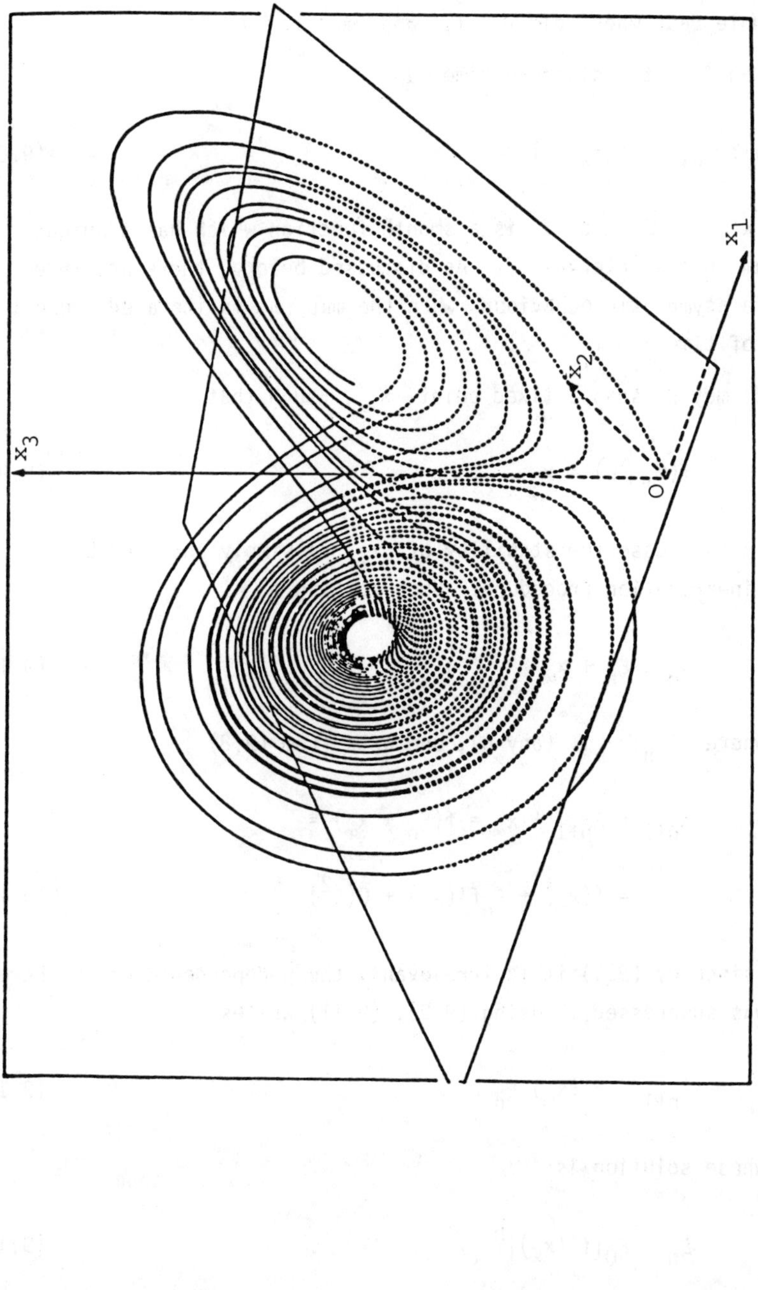

Fig. 30.

set is the point $b = 0$). The approach to chaotic behaviour, one realizes in this way, has interesting universal properties first discovered by M. Feigenbaum and will be briefly discussed later in the simple case when $\dim \mathscr{M} = 1$, and $m = 1$.

Consider the discrete time map

$$\phi : x_{n+1} = f(x_n , \mu) \tag{9.8}$$

where $x \in \mathscr{T} \subset \mathbb{R}$ and f is a smooth function which has a unique extremum in the interval $\mathscr{T}$. As discussed before, there are three possible asymptotic behaviours when the map ϕ is iterated a great number of times:

i) ϕ may possess a fixed point x_∞, such that

$$f(x_\infty , \mu) = x_\infty \quad . \tag{9.9}$$

In this case the stability of x_∞ is simply analyzed by a linearization procedure. Upon setting

$$x_n \equiv \xi_n + x_\infty \tag{9.10}$$

where $|\xi_n| << 1$ (any n), one gets from (9.8)

$$\begin{aligned} x_{n+1} &= \xi_{n+1} + x_\infty = f(\xi_n + x_\infty) = \\ &= f(x_\infty) + \xi_n f'(x_\infty) + 0(\xi^2) \end{aligned} \tag{9.11}$$

[since by (9.9) it is irrelevant, the μ-dependence of f here was suppressed]. Using (9.9), (9.11) writes

$$\xi_{n+1} \approx f'(x_\infty)\, \xi_n \tag{9.12}$$

whose solution is

$$\xi_n \approx \xi_0 [f'(x_\infty)]^n \quad . \tag{9.13}$$

Thus the criterion for stability of ϕ is

$$|f'(x_\infty)| < 1 \quad . \tag{9.14}$$

For stable x_∞, the approach to it is exponential unless $f'(x_\infty) = 0$, a situation in which x_∞ is said to be superstable (the approach to equilibrium is faster than any rate of exponential convergence).

ii) ϕ may have a periodic asymptotic behaviour, i.e. there exist N (possibly $N = \infty$) values of x, say $\{x_\infty^{(r)} \; ; \; r = 0,1,\ldots,N-1\}$, constituting an N-cycle, such that

$$x_\infty^{(r+1)} = f(x_\infty^{(r)}) \quad , \quad r = 0,1,\ldots,N-1 \ (\mathrm{mod}\ N) \quad . \tag{9.15}$$

In other words each $x_\infty^{(r)}$ is a fixed point of the N-th iterate of f, $f^{(N)}$

$$f^{(N)}(x_\infty^{(r)}) = x_\infty^{(r)} \quad , \quad r = 0,1,\ldots,N-1 \tag{9.16}$$

(recalling that

$$f^{(n)}(x) \doteq f\left(f^{(n-1)}(x)\right) \quad , \quad f^{(0)}(x) = x \Big) \quad . \tag{9.17}$$

By an obvious generalization of previous discussion, upon defining

$$\partial f^{(N)}\left(\{x_\infty^{(r)}\}\right) \equiv \prod_{r=0}^{N-1} f'\left(x_\infty^{(r)}\right) \quad , \tag{9.18}$$

the N-cycle is stable if

$$\left[\partial f^{(n)}\left(\{x_\infty^{(r)}\}\right)\right] < 1 \quad . \tag{9.19}$$

An N-cycle is superstable if one of its elements is located at the extremum of f.

iii) ϕ can lead to an aperiodic eventual behaviour, i.e. an infinite sequence of values of x. The latter is stable if any point in the neighbourhood of one of these values, acted on with the mapping (9.8) a great enough number of times, determines a sequence which asymptotically coincides with it,

We intend now to analyze how the nature of the attractors varies as the parameter μ is varied. Notice first that in the proximity of the fixed point $x_\infty^{(\mu)}$, $f(x,\mu)$ can always be expanded in the form

$$f(x,\mu) = -A_\mu\left(x - x_\infty^{(\mu)}\right)^2 + 2A_\mu\left(\bar{x}_\mu - x_\infty^{(\mu)}\right)\left(x - x_\infty^{(\mu)}\right) + \\ + x_\infty^{(\mu)} + 0\left(\left(x - x_\infty^{(\mu)}\right)^3\right) \tag{9.20}$$

where

$$A_\mu = \left.\frac{d^2f}{dx^2}\right|_{x=x_\infty^{(\mu)}} \tag{9.21}$$

and $\bar{x}_\mu$ is the abscissa corresponding to the maximum of f:

$$\left.\frac{df}{dx}\right|_{x=\bar{x}_\mu} = 0 \quad . \tag{9.22}$$

(Notice that (9.20) is consistent with the fixed point Eq. (9.9)). For n sufficently large, the map (9.8) can then be written to one order higher than (9.12),

$$\xi_{n+1} = -A_\mu\xi_n^2 + 2A_\mu\left(\bar{x}_\mu - x_\infty^{(\mu)}\right)\xi_n \quad . \tag{9.23}$$

Introducing the auxiliary coefficients

$$\nu_\mu = 2A_\mu\left(\bar{x}_\mu - x_\infty^{(\mu)}\right)$$

$$a_\mu = \nu_\mu + \frac{1}{\nu_\mu} \quad , \qquad b_\mu = A_\mu\left(\frac{1}{\nu_\mu^3} - 1\right) \quad , \tag{9.24}$$

(9.23) can in turn be written in the form (cf. (6.34)),

$$\xi_{n+1} + \xi_{n-1} = a_\mu \xi_n + b_\mu \xi_n^2 \quad , \tag{9.25}$$

which is easily reduced to a discrete flow in $\mathbb{R}^2$, (we rescale the variables in such a way as to reduce the dependence on μ to a single coefficient $C_\mu = 1/2\ a_\mu$)

$$\psi : \begin{cases} \eta_{n+1} = \xi_n \\ \xi_{n+1} = -\eta_n + 2C_\mu \xi_n + 2\xi_n^2 \end{cases} \quad . \tag{9.26}$$

It is easily checked that (9.26) has a unit Jacobian for any n, and two fixed points: $P_0 = (0\,,0)$; $P_1 = (1 - C_\mu\,, 1 - C_\mu)$. P_0, which is elliptic for $|C_\mu| < 1$, becomes hyperbolic when $C_\mu < -1$. On the other hand, in the latter case a new fixed set appears; a periodic orbit of period 2, constituted by the two points — each of which is fixed under ψ^2 —

$$\left\{P_0^{(r)} \equiv \left(\bar{\eta}_0^{(r)}\,, \bar{\xi}_0^{(r)}\right) = \left(s_\mu + (-)^r|t_\mu|\,, s_\mu - (-)^r|t_\mu|\right) \quad , \right.$$
$$\left. r = 0\,, 1 \ (\mathrm{mod}\ 2)\right\} \tag{9.27}$$

where

$$s_\mu = -\frac{1}{2}(C_\mu + 1) \quad , \quad |t_\mu| = \frac{1}{2}\sqrt{(C_\mu + 1)(C_\mu - 3)} \quad . \tag{9.28}$$

If one evaluates ψ^2 from (9.26)

$$\psi^2 : \begin{cases} \eta_{n+2} = -\eta_n + 2C_\mu \xi_n + 2\xi_n^2 \\ \xi_{n+2} = -\xi_n + 2C_\mu \eta_{n+2} + 2\eta_{n+2}^2 \end{cases} \tag{9.29}$$

one can see that, in terms of the auxiliary variables $x_m \doteq \eta_{2m}$, $y_m \doteq \xi_{2m}$ it can be rewritten for n even (say $n = 2k$) in the form:

$$\tau \equiv \psi^2 : \begin{cases} x_{k+1} = -x_k + 2C_\mu y_k + 2y_k^2 \\ y_{k+1} = -y_k + 2C_\mu x_{k+1} + 2x_{k+1}^2 \end{cases} \tag{9.30}$$

whose fixed points are P_0, P_1, $P_0^{(0)}$, $P_0^{(1)}$. In the neighbourhood of $P_0^{(r)}$, (9.30) is of the form

$$\tau : \begin{cases} \alpha_{k+1} = -\alpha_k + 2\left(C_\mu + 2\bar{\xi}_0^{(r)}\right)\beta_k + 2\beta_k^2 \\ \beta_{k+1} = -\beta_k + 2\left(C_\mu + 2\bar{\eta}_0^{(r)}\right)\alpha_{k+1} + 2\alpha_{k+1}^2 \end{cases} \tag{9.31}$$

where we have set

$$x_k = \bar{\eta}_0^{(r)} + \alpha_k \quad ,$$

$$y_k = \bar{\xi}_0^{(r)} + \beta_k \quad . \tag{9.32}$$

Up to terms of third order in α_k, β_k, Eq. (9.31) writes as

$$\begin{cases} \alpha_{k+1} = -\alpha_k + 2C''\beta_k + 2\beta_k^2 \\ \beta_{k+1} = (4C'C'' - 1)\beta_k - 2C'\alpha_k + 2\alpha_k^2 + \\ \qquad + 4(2C''^2 + C')\beta_k^2 - 8C''\alpha_k\beta_k \end{cases} \tag{9.33}$$

with

$$C' = C_\mu + 2\bar{\eta}_0^{(r)}$$

$$C'' = C_\mu + 2\bar{\xi}_0^{(r)} \quad . \tag{9.34}$$

Finally the new change of variables

$$\alpha_k \doteqdot X_k$$

$$-\alpha_k + 2C''\beta_k + 2\beta_k^2 \doteqdot Y_k \tag{9.35}$$

leads to writing (9.33) in the form (once more up to third order terms in X_k, Y_k),

$$\begin{cases} X_{k+1} = Y_k \\ Y_{k+1} = -X_k + 2(2C'C'' - 1)Y_k + 4C''Y_k^2 + R_k \end{cases} \quad . \tag{9.36}$$

The remainder $R_k = \frac{1}{C''}\left\{[(2C'C'' - 1)Y_k - X_k]^2 + 4C'^2C''^2Y_k^2\right\}$ is indeed of higher order in X_k, Y_k, and can be neglected. Upon rescaling the variables: $\tilde{\xi}_n \doteqdot 2C''Y_n$, $\tilde{\eta}_n \doteqdot 2C''X_n$, (9.36) writes:

$$\begin{cases} \tilde{\eta}_{k+1} = \tilde{\xi}_k \\ \tilde{\xi}_{k+1} = -\tilde{\eta}_k + 2\tilde{C}_\mu\tilde{\xi}_k + 2\tilde{\xi}_k^2 \end{cases} \tag{9.37}$$

which is indeed identical with (9.26), with C_μ replaced by

$$\tilde{C}_\mu = 2C'C'' - 1 = -2C_\mu^2 + 4C_\mu + 7 \quad . \tag{9.38}$$

The entire procedure can now be repeated, leading to a succession of bifurcations with doubling of the period, controlled by the values of the parameter C_μ (or its iterates according to (9.38)). Equation (9.38) can be written recursively as

$$C^{(k)} = -2\left(C^{(k+1)}\right)^2 + 4C^{(k+1)} + 7 \tag{9.39}$$

with $C^{(0)} = C_\mu$, $C^{(1)} = \tilde{C}_\mu$.

The sequence has an asymptotic value, which is a fixed point,

$$C^{(\infty)} = \frac{3 - \sqrt{65}}{4} \tag{9.40}$$

reached at exponential rate

$$C^{(k)} = C^{(\infty)} + a\delta^{-k} \quad , \quad \delta = \frac{C^{(k)} - C^{(k-1)}}{C^{(k+1)} - C^{(k)}} \quad . \tag{9.41}$$

The constant δ can be calculated substituting (9.41) in (9.30) and comparing the lowest order terms

$$\delta = - 4C^{(\infty)} + 4 = 1 + \sqrt{65} \quad . \tag{9.42}$$

The numerical studies confirm that this value of δ is indeed universal and holds for all "conservative" (i.e. area preserving in the phase plane) systems. The same kind of universality can also be found in "dissipative" systems (namely systems for which the Jacobian is $\neq 1$, corresponding to area contracting or area expanding maps of the phase plane onto itself), with a different value of the constant δ.

In order to find the latter, consider the slightly modified form of (9.26)

$$\psi_B : \begin{cases} \eta_{n+1} = \xi_n \\ \\ \xi_{n+1} = B\eta_n + 2C_\mu \xi_n + 2\xi_n^2 \end{cases} \tag{9.43}$$

where B = Jacobian (ψ_B). If one assumes $|B| < 1$ (if $|B| > 1$, one should just consider the time-reversed mapping in the variables $\tilde{\xi}_n \doteqdot \xi_{-n}$, $\tilde{\eta}_n \doteqdot \eta_{-n}$), ψ_B is dissipative, area contracting. By introducing the auxiliary constant

$$D_\mu \doteqdot C_\mu (B - 1 + C_\mu) \tag{9.44}$$

and upon performing the change of variables

$$\xi_n = -\frac{1}{2}(D_\mu X_n + C_\mu)$$

$$\eta_n = -\frac{1}{2}(D_\mu Y_n + C_\mu) \qquad , \qquad (9.45)$$

(9.43) could write

$$\psi_B : \begin{cases} Y_{n+1} = X_n \\ \\ X_{n+1} = BY_n + 1 - D_\mu X_n^2 \end{cases} \qquad (9.46)$$

or (Hénon dissipative mapping)

$$X_{n+1} - BX_{n-1} = 1 - D_\mu X_n^2 \qquad . \qquad (9.47)$$

The whole discussion can now be repeated with minor changes for any of the above forms. Confining the attention on the form (9.43) written as

$$X_{n+1} - BX_{n-1} = 2C_\mu X_n + 2X_n^2 \qquad (9.48)$$

one immediately notices that two obvious period-1 orbits are, upon setting once more

$$X_n = \hat{X} + x_n \qquad , \qquad |x_n| << 1 \qquad (9.49)$$

given by $\hat{X} = 0$ and $\hat{X} = \frac{1}{2}(1 - B - 2C_\mu)$. Their variational equation [up to $O(x_n^2)$] is

$$x_{n+1} - Bx_{n-1} = 2(C_\mu + 2\hat{X})x_n \qquad , \qquad (9.50)$$

whose solutions

$$x_n = \alpha\lambda_1^n + \beta\lambda_2^n \quad \text{and} \tag{9.51}$$

where $\lambda_{1,2}$ are roots of the equation

$$\lambda^2 - 2(C_\mu + 2\hat{X})\lambda - B = 0 \quad , \tag{9.52}$$

remain bounded for all n's if and only if $|\lambda_i| < 1$, $i = 1,2$. In (9.51)

$$\alpha = \frac{x_1 - x_0\lambda_2}{\lambda_1 - \lambda_2} \quad , \quad \beta = \frac{x_0\lambda_1 - x_1}{\lambda_1 - \lambda_2} \quad . \tag{9.53}$$

Considering for simplicity the case $\hat{X} = 0$, the origin is then stable if

$$|C_\mu| \leq \frac{1}{2}(1 - B) \quad . \tag{9.54}$$

It is also straightforward to check that a period-2 orbit exists, given by the fixed points $\hat{X}_s$, $s = 0,1 \pmod 2$ of the mapping (9.48) squared:

$$\hat{X}_s = -\frac{1}{2}\left[C_\mu + \frac{1}{2}(1-B)\right] + \\ + (-)^s \frac{1}{2}\left[C_\mu + \frac{1}{2}(1-B)\right]^{\frac{1}{2}}\left[C_\mu - \frac{3}{2}(1-B)\right]^{\frac{1}{2}} \tag{9.55}$$

which are real for

$$C_\mu \leq -\frac{1}{2}(1-B) \quad \text{and} \quad C_\mu \geq \frac{3}{2}(1-B) \quad . \tag{9.56}$$

Notice that the former inequality gives exactly the limit at which the origin turns instable.

Denoting once more by $C^{(k)}$ the value of C_μ at which a stable period-2^k orbit is created, (9.54) and (9.56) read

$$C^{(0)} = \frac{1}{2}(1-B) = -C^{(1)} \quad . \tag{9.57}$$

Substituting (9.49) into (9.48), the original mapping is transformed into

$$x_{n+1} - Bx_{n-1} = 2(C_\mu + 2\hat{X})x_n + 2x_n^2 \quad . \tag{9.58}$$

After writing now (9.58) for $n = 2t + 1$, $n = 2t - 1$ and $n = 2t$, one adds the second, multiplied by $-B$ to the first, and substitutes in the resulting expression the third — in such a way as to eliminate the linear odd-step terms. The final result is

$$\begin{aligned} x_{2t+2} + B^2 x_{2t-2} &= 2C'_\mu x_{2t} + 2C''_\mu x_{2t}^2 + \\ &\quad + 2(x_{2t+1}^2 - Bx_{2t-1}^2) \end{aligned} \tag{9.59}$$

where

$$\begin{aligned} C'_\mu &= B + 2D_\mu^{(0)} D_\mu^{(1)} \\ C''_\mu &= 2D_\mu^{(1)} \end{aligned} \tag{9.60}$$

with

$$D_\mu^{(s)} = C_\mu + 2\hat{X}_s \quad , \qquad s = 0,1 \pmod 2 \quad . \tag{9.61}$$

Setting, as in (9.29) - (9.30), $x_{2m} \doteq \xi_m \cdot a$, with

$$a = \left[C''_\mu + \frac{4}{(1-B)} \frac{(C'_\mu - B)^2}{C''^2_\mu} \right]^{-1} \tag{9.62}$$

(9.59) writes

$$\xi_{t+1} - B'\xi_{t-1} = 2C'_\mu \xi_t + 2\xi_t^2 + \tilde{R}_t \tag{9.63}$$

where

$$B' = - B^2 \tag{9.64}$$

and the remainder $\tilde{R}_t$ is of higher order [compare Eq. (9.36)]. Neglecting $\tilde{R}_t$, (9.63) has the same form as (9.48) with both B and C_μ rescaled according to (9.64) and (9.60) respectively.

Hence there is again a period doubling bifurcation at $C'_\mu = -\frac{1}{2}(1 - B')$. Observing that the first equation in (9.60) reads explicitly, making use of (9.61) and (9.55),

$$C'_\mu = - 2C_\mu^2 + 2(1 - B)C_\mu + 2B^2 - 3B + 2 \tag{9.65}$$

and recalling (9.64), the new critical value of C_μ is thus given by

$$C^{(2)} = \frac{1}{2}\left\{1 - B - \sqrt{6B^2 - 8B + 6}\right\} . \tag{9.66}$$

[Notice how, letting $B = -1$, (9.66) gives $C^{(2)} = 1 - \sqrt{5}$ which coincides with the solution of (9.39) corresponding to the initial conditions (9.57), $C^{(0)} = -C^{(1)} = 1$]. Since (9.64) implies for the k-th iterate $B^{(k)}$ of the coefficient B,

$$B^{(k)} = - B^{2^k} \tag{9.67}$$

and since by the assumption that $|B| < 1$, $B^{(k)} \to 0$ as $k \to \infty$, the dissipative mapping is in this limit equivalent to a first difference mapping

$$X_{n+1} \approx 2CX_n + 2X_n^2 . \tag{9.68}$$

Asymptotically the iteration of (9.64) and (9.65) leads then simply to the equation (to be compared with (9.39), valid for the conservative case, $B = -1$),

$$C^{(k)} \cong -2[C^{(k+1)}]^2 + 2C^{(k+1)} + 2 \qquad (9.69)$$

whereby the limiting value

$$C^{(\infty)}_{\text{diss.}} = \frac{1}{4}(1 - \sqrt{17}) \qquad (9.70)$$

can be derived. Even in this case the rate of convergence is a universal constant:

$$C^{(k)} = C^{(\infty)}_{\text{diss.}} + \varepsilon\delta_0^{-k} \quad . \qquad (9.71)$$

The parameter

$$\delta_0 = -4C^{(\infty)}_{\text{diss.}} + 2 = 1 + \sqrt{17} \qquad (9.72)$$

is referred to as Feigenbaum's constant. Notice that the value (9.72) is slightly overestimated with respect to the exact value (obtained numerically by Feigenbaum)

$$\delta_F = 4.6692016091029909 \ldots \qquad (9.73)$$

because of the truncation performed on the mapping (9.63). Also the scaling parameter a^{-1} of (9.62) has a universal asymptotic value:

$$\alpha \doteq -a^{-1} = 2.502907875095892884 \ldots \qquad (9.74)$$

A better estimate of δ_F can be obtained in the following way. Consider the k-step composed map of (9.8):

$$x_{t+k} = f^{(k)}(x_t, \mu) \quad . \qquad (9.75)$$

A cycle of period k touches the points $\bar{x}_k^{(r)}$, $r = 1,\ldots,k$, which are fixed points of $f^{(k)}$, solutions of

$$\bar{x}_k^{(r)} = f^{(k)}\left(\bar{x}_k^{(r)}, \mu\right) \quad , \quad r = 1,\ldots,k \quad . \tag{9.76}$$

Let

$$\lambda_k^{(r)}(\mu) = \frac{d}{dx} f^{(k)}(x, \mu)\Bigg|_{x=\bar{x}_k^{(r)}} \quad , \quad r = 1,\ldots,k \tag{9.77}$$

be the stability determining (see Eqs. (9.18), (9.19)) slope of $f^{(k)}$ at these points, considered as a function of the parameter μ. If $|\lambda_k^{(r)}| < 1$, then the point is stable and becomes unstable when $|\lambda_k^{(r)}| > 1$.

The 2k-composed map $f^{(2k)}$ then undergoes a bifurcation to produce two new fixed points of period $2k$, $\bar{x}_{2k}^{(r')}$ for each $\bar{x}_k^{(r)}$ ($r = 1,\ldots,k$; $r' = 1,\ldots,2k$). The Feigenbaum ratio δ (see definition (9.41)) can thus equivalently be defined as

$$\delta_F = \lim_{k\to\infty} \frac{\Delta\mu(k)}{\Delta\mu(2k)} \tag{9.78}$$

where $\Delta\mu(k)$ denotes the variation of the parameter μ as λ_k sweeps the range from -1 to $+1$. Assuming that both $f^{(k)}$ and $f^{(2k)}$ are well approximated by cubic relations in the domain shown in Fig. 31, and further requiring that $\bar{x}_{2k}^{(q)} - \bar{x}_k^{(r)} \cong \bar{x}_k^{(r)} - \bar{x}_{2k}^{(s)}$ (i.e. the two new fixed points are symmetrically placed around the previous ones; this assumption is not strictly necessary but makes calculations easier), one can easily check that they must write

$$f^{(k)}(x) = \bar{x}_k + \lambda_k(x - \bar{x}_k) + \frac{(1-\lambda_k^2)}{\lambda_k(1+\lambda_k^2)}\,\frac{1}{(\bar{x}_{2k}-\bar{x}_k)^2}\,(x-\bar{x}_k)^3 + $$
$$+ O\left((x-\bar{x}_k)^4\right) \quad , \tag{9.79}$$

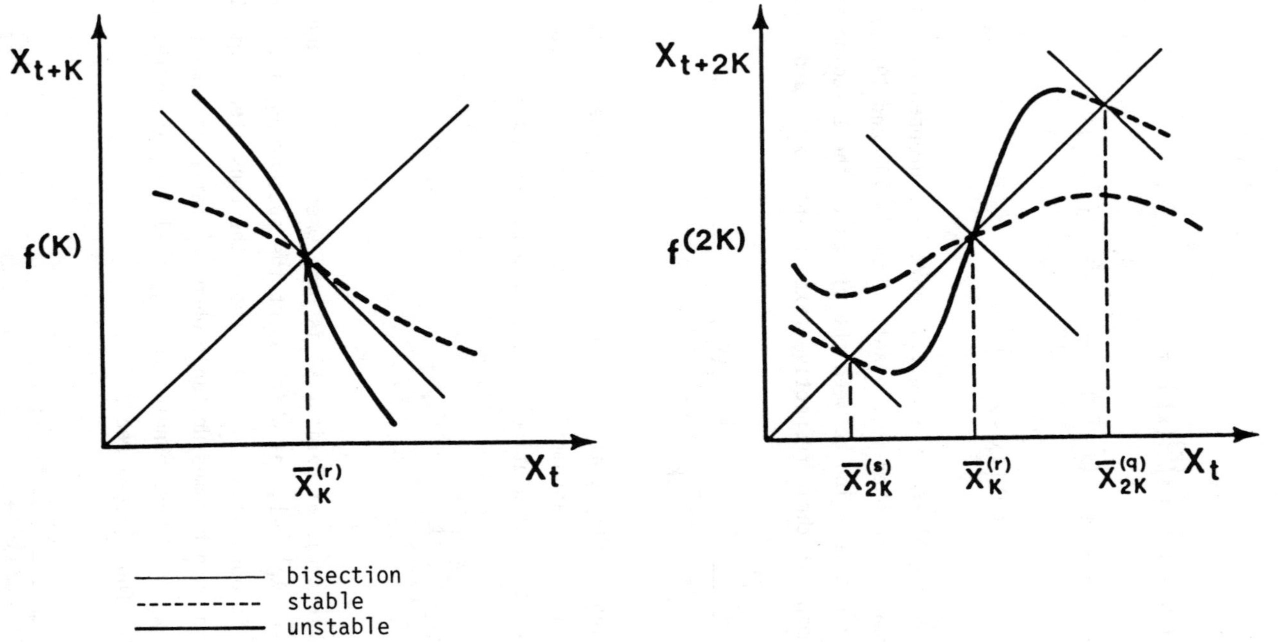

X_{t+K}
$f^{(K)}$
$\overline{X}_K^{(r)}$
X_t
X_{t+2K}
$f^{(2K)}$
$\overline{X}_{2K}^{(s)}$
$\overline{X}_K^{(r)}$
$\overline{X}_{2K}^{(q)}$
X_t
bisection
stable
unstable

Fig. 31.

and

$$f^{(2k)}(x) = f^{(k)}\left(f^{(k)}(x)\right) =$$

$$= \bar{x}_k + \lambda_k^2(x - \bar{x}_k) + \frac{(1 - \lambda_k^2)}{(\bar{x}_{2k} - \bar{x}_k)^2}(x - \bar{x}_k)^3 +$$

$$+ 0\left((x - \bar{x}_k)^4\right) \tag{9.80}$$

where the upper indices $r, s, q, \ldots$ have been dropped because the hypotheses made make them irrelevant. From (9.79) and (9.80) it is easy to derive a relation between the slopes of the 2k-composed and k-composed maps at their respective fixed points $\bar{x}_{2k}$ and $\bar{x}_k$:

$$\lambda_{2k} = \left.\frac{df^{(2k)}}{dx}\right|_{x=\bar{x}_{2k}} = 3 - 2\lambda_k^2 + \ldots \tag{9.81}$$

Assuming finally that the slope λ depends linearly on the parameter μ (also this assumption is not strictly necessary but for made simplicity of the calculation), (9.78) can be rewritten

$$\delta = \lim_{k\to\infty} \frac{(\Delta\lambda)_k}{(\Delta\lambda)_{2k}} \quad . \tag{9.82}$$

Now the stable cycle of period k is born when $\lambda_k = +1$ and becomes unstable when $\lambda_k = -1$, at which point the succeeding stable cycle of period $2k$ is born with $\lambda_{2k} = 1$ (check the consistency of (9.81)). The latter in turn becomes unstable when λ_k attains some (negative) value, say $\lambda_k^{(c)}$, corresponding to $\lambda_{2k} = -1$. Equation (9.81) gives $\lambda_k^{(c)} = -\sqrt{2}$. Thus (9.82) reads

$$\delta = \lim_{k\to\infty} -\frac{2}{\lambda_k^{(c)} + 1} \approx 2(1 + \sqrt{2}) \quad , \tag{9.83}$$

a value which constitutes a much better approximation of the value (9.73) of δ_F than that of (9.72).

Returning now to the general map ϕ of (9.8), the lesson that is to be learnt from the two examples thoroughly analysed so far is that the process repeatedly adopted: functional composition followed by magnification (determined by the fixed point condition), has a limit. Upon defining such a limit as

$$g(x) = \lim_{n\to\infty} (-\alpha)^n f^{(2^n)}\left(\frac{x}{(-\alpha)^n}, \mu_{n+1}\right) \quad , \tag{9.84}$$

where $f^{(k)}$ denotes the functional composition:

$$f^{(k)} = f \; f^{(k-1)} \quad , \quad f^{(0)} = x \quad , \quad f^{(1)} = f \tag{9.85}$$

and μ_k is the value of the parameter μ corresponding to $C_\mu \equiv C^{(k)}$. We know now that g must be self-similar by composition i.e.

$$g(x) = -\alpha \, g\left(g(-x/\alpha)\right) \quad . \tag{9.86}$$

Notice that Eq. (9.86) determines α as well as $g(x)$. Indeed the theory defined by it has no absolute scale in it, in that it is dilation invariant. The scale should be fixed conventionally, e.g. setting $g(0) = 1$, in which case $\alpha = -1/g(1)$. Naturally it also determines δ_F. In fact, let us expand f as a function of the parameter in the neighbourhood of μ_∞:

$$f(x, \mu) = f(x, \mu_\infty) + (\mu - \mu_\infty) f_\mu(x, \mu_\infty) + 0\left([\mu - \mu_\infty]^2\right)$$

$$\equiv G_0(x) + (\mu - \mu_\infty) H_0(x) + 0\left([\mu - \mu_\infty]^2\right) \tag{9.87}$$

(f_μ of course denotes $\partial f/\partial\mu$). Define analogously

$$(-\alpha)^n f^{(2^n)}\left(\frac{x}{(-\alpha)^n}, \mu_{n+1}\right) \equiv G_n(x) + (\mu_{n+1} - \mu_\infty)H_n(x) + $$
$$+ 0\left([\mu_{n+1} - \mu_\infty]^2\right) \quad . \qquad (9.88)$$

Iteration of (9.88) straightforwardly gives the recurrence relation for the auxiliary sets of functions $\{G_n\}, \{H_n\}$:

$$G_{n+1}(x) = -\alpha\, G_n\left[G_n\left(-\frac{x}{\alpha}\right)\right] \qquad (9.89)$$

$$H_{n+1}(x) = -\alpha\left\{H_n\left[G_n\left(-\frac{x}{\alpha}\right)\right] + G_n'\left[G_n\left(-\frac{x}{\alpha}\right)\right]H_n\left(-\frac{x}{\alpha}\right)\right\} \quad . \qquad (9.90)$$

Comparing (9.89) with (9.86) suggests that as $n \to \infty$, $G_n(x) \to g(x)$, so that the limit of large n of Eq. (9.90) can be written

$$H_{n+1}(x) \sim -\alpha\left[H_n\left(g\left(-\frac{x}{\alpha}\right)\right) + g'\left(g\left(-\frac{x}{\alpha}\right)\right)H_n\left(-\frac{x}{\alpha}\right)\right] \quad . \qquad (9.91)$$

On the other hand, comparing (9.88) with (9.84) suggests that as $n \to \infty$, H_n should depend on n exponentially, with rate δ. Therefore one sets

$$H_n(x) = c_f h(x)\delta^n \qquad (9.92)$$

where c_f is a constant (depending on the explicit form of f). $h(x)$ is then the eigenfunction, corresponding to eigenvalue δ, of the functional equation — obtained from inserting (9.92) into (9.91) —

$$-\alpha\left\{h\left(g\left(-\frac{x}{\alpha}\right)\right) + g'\left(g\left(-\frac{x}{\alpha}\right)\right)h\left(-\frac{x}{\alpha}\right)\right\} = \delta h(x) \quad . \qquad (9.93)$$

The exact numerical values (9.73), (9.74) were obtained without any truncation by solving (9.86) and (9.93).

One more general property can be derived for the mapping ϕ in (9.8) when $f(x)$ is not monotonic. It is expressed by the noteworthy theorem, due to Šarkovskiĭ, whose statement is given below.

Write the positive integers in the order (denoted by the symbol $\triangleleft$):

$$3 \triangleleft 5 \triangleleft 7 \triangleleft \dots \triangleleft 2\cdot 3 \triangleleft 2\cdot 5 \triangleleft 2\cdot 7 \triangleleft \dots \triangleleft 2^2\cdot 3 \triangleleft 2^2\cdot 5 \triangleleft 2^2\cdot 7 \triangleleft$$

$$\dots \triangleleft 2^3 \triangleleft 2^2 \triangleleft 2 \triangleleft 1 \tag{9.94}$$

namely:

i) ascending odds >1 come first;
ii) descending powers of 2 come last;
iii) each sequence of ascending odds >1 times a fixed power of 2 comes between, with lower powers of 2 coming earlier;

then if $n \triangleleft k$, and ϕ has a periodic point of period n, then ϕ has also a periodic point of period k. In other words the existence of a periodic point with period n implies the existence of periodic points with periods m for each and all m's following n in the ordered sequence (9.94).

The proof of the above theorem requires a few preliminary notions that will be briefly reviewed. Let $\mathscr{A}_s = \{\mathscr{I}_1,\dots,\mathscr{I}_s\}$, $s \geq 1$, be a partition of the interval $\mathscr{I}$ (see (9.8)) into s sub-intervals, i.e. a family of closed intervals such that

$$\mathscr{I}_1 \cup \dots \cup \mathscr{I}_s = \mathscr{I} \tag{9.95}$$

and, for $i \neq j$, $\mathscr{I}_i \cap \mathscr{I}_j$ consists of at most one point. The interval $\mathscr{I}$ is said to "ϕ-cover" the domain $\mathscr{J}$ n-times, if there exists a partition $\mathscr{A}_n$ of $\mathscr{I}$, (or, for $n=1$, a sub-interval $\mathscr{I}_1 \subseteq \mathscr{I}$) such that

$$f(\mathscr{I}_i) = \mathscr{J} \quad \text{for} \quad i = 1,\dots,n \quad . \tag{9.96}$$

Obviously, if $\mathscr{I}$ ϕ-covers $\mathscr{J}$ and $\mathscr{K} \subset \mathscr{J}$, then $\mathscr{I}$ ϕ-covers $\mathscr{K}$ as well. Upon normalizing the function f so that ϕ maps $\mathscr{I}$ onto itself, one can then define an $\mathscr{A}$-graph $\mathscr{G}_n$ of ϕ in the following way:

i) the n vertices of $\mathscr{G}_n$ are disjoint points on the plane, each associated to a sub-interval $\mathscr{I}_i$, $i = 1,\dots,n$ of $\mathscr{A}_n$;

ii) such vertices are connected by oriented edges according to the rule that if $\mathscr{I}_i$ ϕ-covers $\mathscr{I}_j$ $(i, j = 1,\dots,n)$ k times, but not $(k+1)$ times, then there are k edges pointing from the vertex associated to $\mathscr{I}_i$ to that associated with $\mathscr{I}_j$.

Figure 32 below illustrates with an example $(n = 3)$ the construction of $\mathscr{G}_n$.

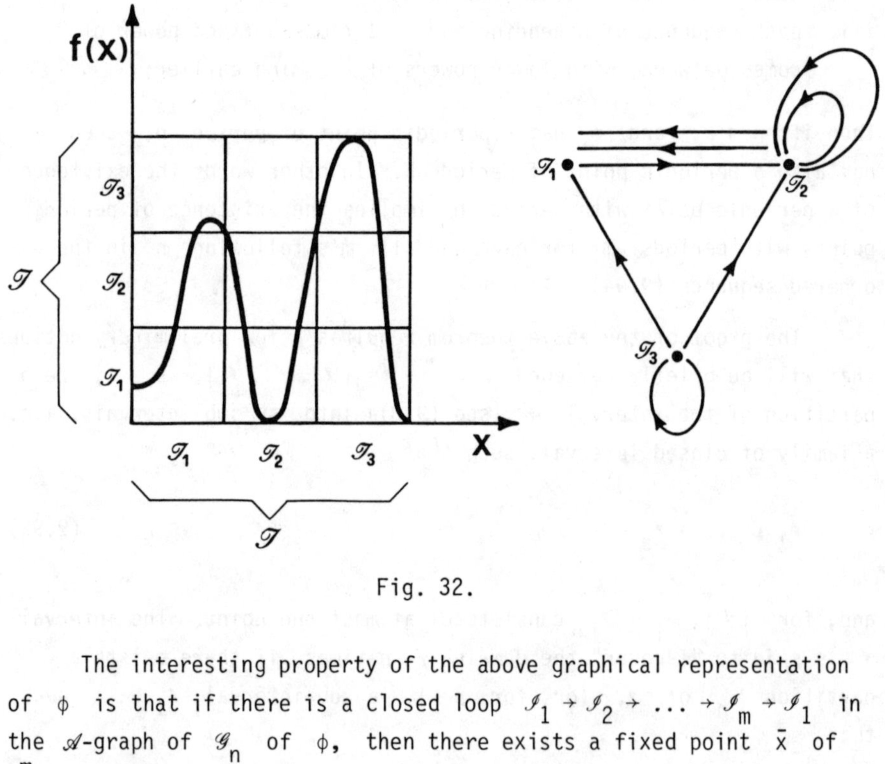

Fig. 32.

The interesting property of the above graphical representation of ϕ is that if there is a closed loop $\mathscr{I}_1 \to \mathscr{I}_2 \to \dots \to \mathscr{I}_m \to \mathscr{I}_1$ in the $\mathscr{A}$-graph of $\mathscr{G}_n$ of ϕ, then there exists a fixed point $\bar{x}$ of ϕ^m, such that

$$f^{(k-1)}(\bar{x}) \in \mathscr{I}_k \quad , \quad \text{for } k = 1,\dots,m \quad . \tag{9.97}$$

Indeed the hypothesis implies that there are intervals $\mathcal{J}_1,\ldots,\mathcal{J}_m$ such that for each k, $\mathcal{J}_k \subset \mathcal{I}_k$, and $f(\mathcal{J}_k) = \mathcal{J}_{k+1}$, with $\mathcal{I}_{m+1} \equiv \mathcal{I}_1$. But then $f^{(m)}(\mathcal{J}_1) = \mathcal{I}_1$ i.e. $\mathcal{I}_1$ ϕ-covers $\mathcal{I}_1$. Thus there are two points, say $v, w \in \mathcal{J}_1$ such that $f^{(m)}(v), f^{(m)}(w)$ are the end-points of $\mathcal{I}_1$. Letting $\mathcal{L} = [v, w]$, $\mathcal{L} \subseteq \mathcal{J}_1$, ϕ^m has a fixed point $\bar{x}$ in $\mathcal{L}$ by continuity; further $\bar{x} \in \mathcal{I}_1$, $f(\bar{x}) \in \mathcal{I}_2,\ldots,f^{(m-1)}(\bar{x}) \in \mathcal{I}_m$. Now, if ϕ has a fixed point $\bar{y}$ of period n, and $n > 1$ is odd and moreover ϕ has no periodic points of odd periods smaller than n and larger than 1, the $\mathcal{A}$-graph of ϕ must contain a subgraph $\mathcal{S}$ of the following form,

$\mathcal{S}$:

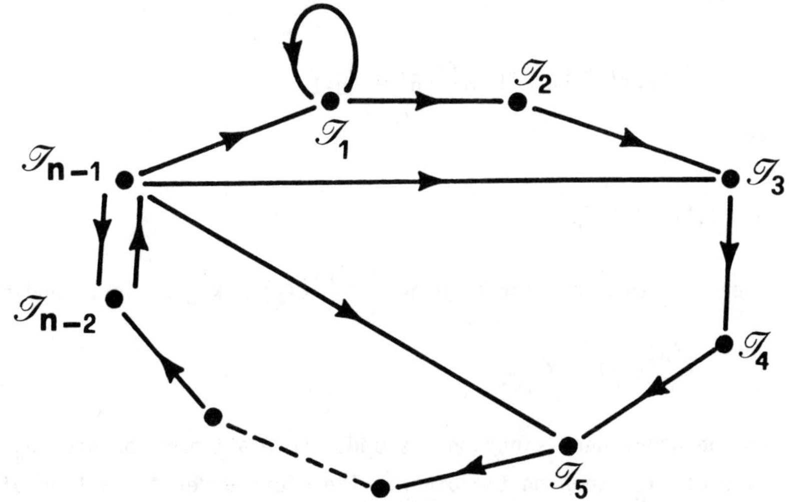

where there are oriented edges from $\mathcal{I}_{n-1}$ to all odd labelled vertices.

In order to check the latter statement, consider the finite set of points $\mathcal{P} = \{p_i = f^{(i-1)}(\bar{y});\ i = 1,\ldots,n\}$. $\mathcal{P}$ is the orbit of $\bar{y}(\equiv p_1)$. Define then $\mathcal{J}$ to be the interval $\mathcal{J} = \left[\min_{1\le i\le n} p_i,\ \max_{1\le i\le n} p_i\right]$, (in present case f has been assumed to be rescaled so that $\mathcal{J} = \mathcal{I}$), and let $\mathscr{A}_{n-1}$ be the partition of $\mathscr{J}$ by the elements of the orbit $\mathscr{P}$:

$$\mathscr{I}_k = [\bar{p}_k , \bar{p}_{k+1}] \quad , \quad k = 1,\ldots,n-1 \tag{9.98}$$

where $\{\bar{p}_i, \ i = 1,\ldots,n\}$ is the reordering of $\{p_i, \ i = 1,\ldots,n\}$ $\bar{p}_i = p_{j_i}$ for some j_i ; $i,j_i = 1,\ldots,n$ such that $\min\limits_{1\le i\le n} p_i = \bar{p}_1 <$ $\bar{p}_2 < \ldots < \bar{p}_{n-1} < \bar{p}_n = \max\limits_{1\le i\le n} p_i$.

Since obviously $f\left(\max\limits_{1\le i\le n} p_i\right) \le \max\limits_{1\le i\le n} p_i$, there is an element, say $\mathscr{I}_{\bar{n}}$ of $\mathscr{A}_{n-1}$, whose left end point $\bar{p}_{\bar{n}-1}$ is the largest element y of the orbit of $\bar{y}$ such that $f(y) > y$. In other words

$$f(\bar{p}_{\bar{n}-1}) \ge \bar{p}_{\bar{n}} \quad , \quad f(\bar{p}_{\bar{n}}) \le \bar{p}_{\bar{n}-1} \tag{9.99}$$

and hence

$$f(\mathscr{I}_{\bar{n}}) \supset \mathscr{I}_{\bar{n}} \quad . \tag{9.100}$$

There follows that the sequence $f^{(k)}(\mathscr{I}_{\bar{n}})$, $k \ge 0$ is ascending, and

$$f^{(n)}(\mathscr{I}_{\bar{n}}) \supset \mathscr{J} \quad . \tag{9.101}$$

On the other hand since n is odd, there are more points p_j on one side of $\mathscr{I}_{\bar{n}}$ than on the other. Therefore under the action of ϕ some of them remain on the same side, others change. Consequently for some element $\mathscr{I}_{\bar{m}}$ of $\mathscr{A}$,

$$f(\mathscr{I}_{\bar{m}}) \supseteq \mathscr{I}_{\bar{n}} \quad . \tag{9.102}$$

There must therefore be a subgraph $\mathscr{B}$ of the $\mathscr{A}$-graph of ϕ of the form

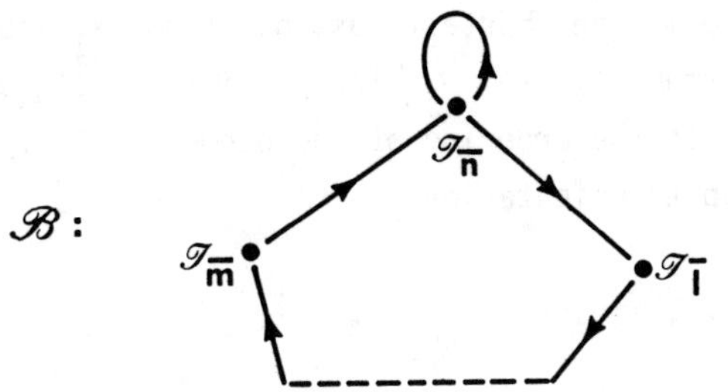

with say, v vertices.

Let us relabel the intervals $\mathcal{I}_k$, $k = 1,\dots,n$ by a permutation π of $\{1,\dots,n\}$ such that $\bar{n} \to 1$, $\bar{\ell} \to 2,\dots,\bar{m} \to v$ (ordered in such a way that the loop is the shortest one possible; with all the remaining labels invariant): $\bar{\mathcal{I}}_k = \mathcal{I}_{\pi(k)}$, $k = 1,\dots,n$. If $v < n-1$, then either one of the loops

$$\bar{\mathcal{I}}_1 \to \bar{\mathcal{I}}_2 \to \dots \to \bar{\mathcal{I}}_v \to \bar{\mathcal{I}}_1$$

or

$$\bar{\mathcal{I}}_1 \to \bar{\mathcal{I}}_2 \to \dots \to \bar{\mathcal{I}}_v \to \bar{\mathcal{I}}_1 \circlearrowright$$

should give a fixed point $\bar{x} \in \bar{\mathcal{I}}_1$ of ϕ^m for some $1 < m < n$, m odd, thus contradicting the hypothesis. Then it must be $v = n-1$ and the subgraph $\mathcal{B}$ must have the form

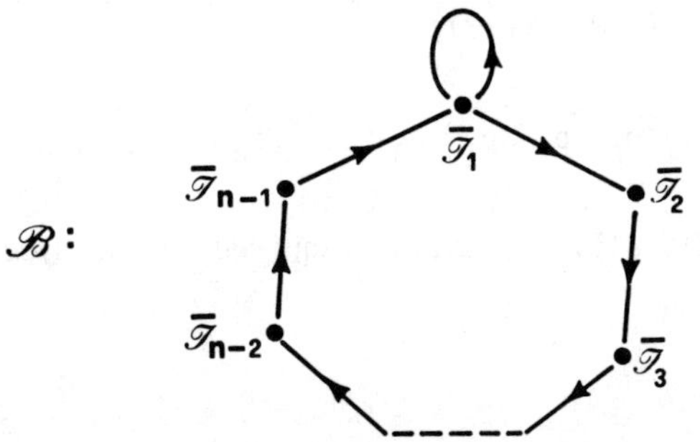

where, since the loop is the shortest possible from $\bar{\mathscr{I}}_1$ to itself, there are no arrows from $\bar{\mathscr{I}}_i$ to $\bar{\mathscr{I}}_j$ for $j > i + 1$. It can now be shown by induction that the ordering of the elements of $\mathscr{A}_n$ on the real line must be (up to orientation):

$$\bar{\mathscr{I}}_{n-1} , \bar{\mathscr{I}}_{n-3} ,\dots, \bar{\mathscr{I}}_2 , \bar{\mathscr{I}}_1 , \bar{\mathscr{I}}_3 ,\dots, \bar{\mathscr{I}}_{n-4} , \bar{\mathscr{I}}_{n-2} \quad . \tag{9.103}$$

In fact, setting up a periodic orbit of odd period $n > 1$ according to

$$f(p_i) = p_{i+1}, \quad (i \bmod n)$$

$$p_{n-1} < \dots < p_4 < p_2 < p_1 < p_3 \dots < p_{n-2} < p_n \quad , \tag{9.104}$$

(and extending f linearly in the intervals between) can be done only in the way illustrated in Fig. 33 (for $n = 5$).

The induction precedure can be set up noticing that for $n = 3$ one has the configuration shown in Fig. 34 and that if the partition (9.103) holds for a given n, it holds for $n + 2$ only provided one adds an extra interval $(\bar{\mathscr{I}}_{n+1})$ at the extreme left and one $(\bar{\mathscr{I}}_{n+2})$ at the extreme right, because one wants:

i) the common end-point of $\bar{\mathscr{I}}_{n-1}$ and $\bar{\mathscr{I}}_{n+1}$ to be mapped onto the point p_{n+2} (common end-point of $\bar{\mathscr{I}}_{n+2}$ and $\mathscr{I}$) — as, by assumption, $f(\bar{\mathscr{I}}_{n-3} \cap \bar{\mathscr{I}}_{n-1}) = p_n$ — in order that the new sequence of points $p_1,\dots,p_{n+1} , p_{n+2}$ satisfies (9.104):

$$f(p_i) = p_{i+1} \ , \quad (i \bmod (n+2)) \quad ,$$

$$p_{n+1} < p_{n-1} < \dots < p_4 < p_2 < p_1 < p_3 < \dots < p_n < p_{n+2} \quad , \tag{9.105}$$

i.e. it is cyclically closed, as required by the hypothesis.

ii)

$$f(\bar{\mathscr{I}}_{n+1}) \supseteq \bar{\mathscr{I}}_1 \tag{9.106}$$

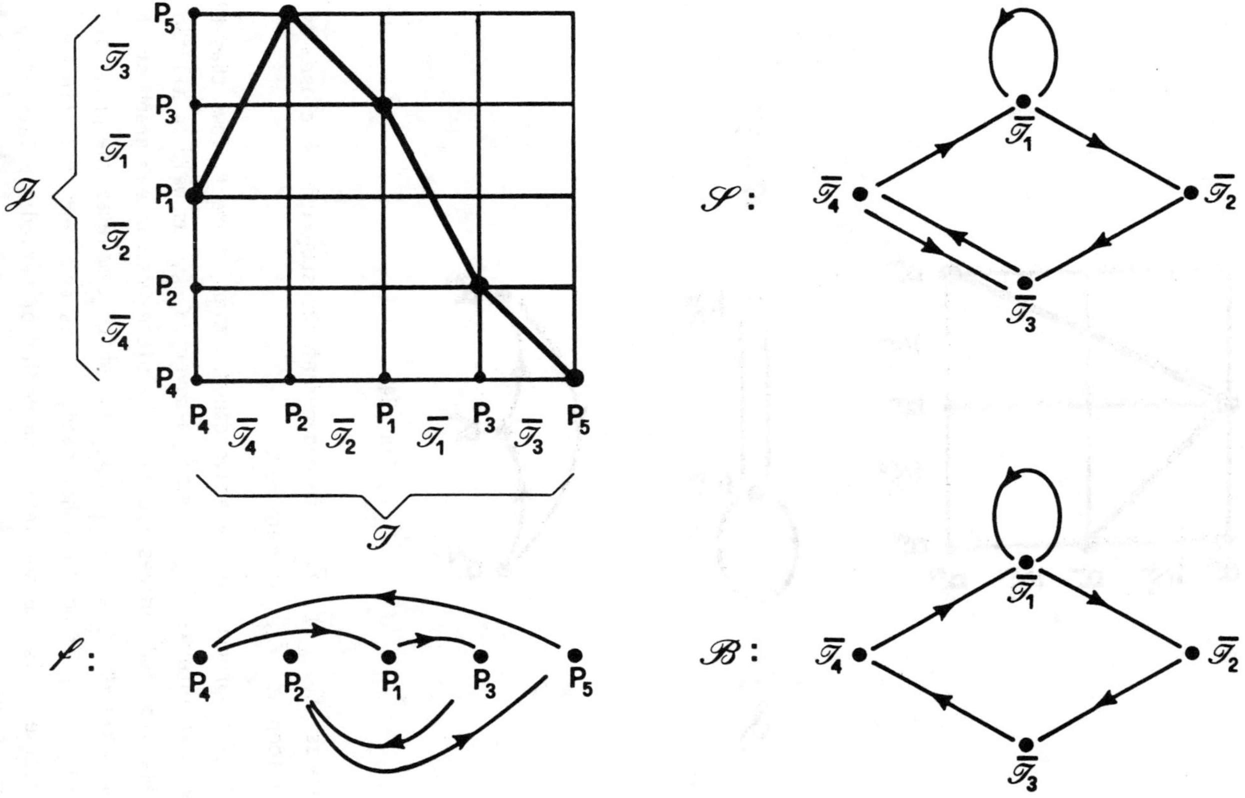

Fig. 33.

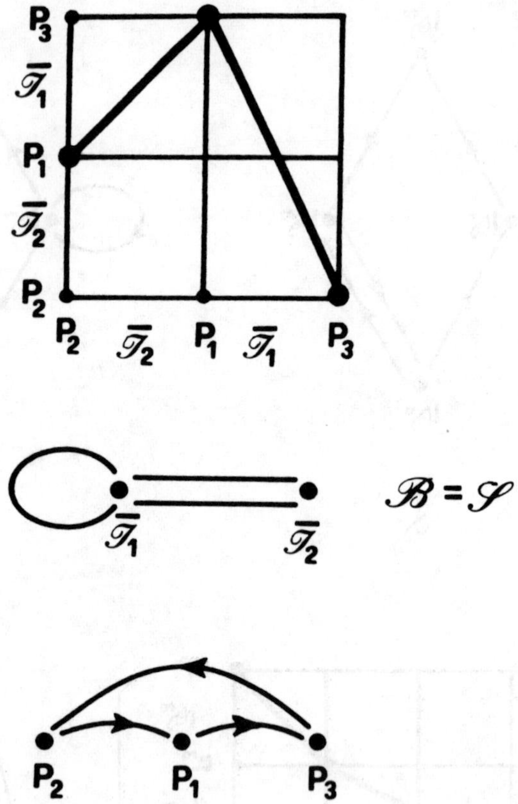

Fig. 34.

— as $f(\bar{\mathcal{I}}_{n-1}) \supseteq \bar{\mathcal{I}}_1$ — in order that the subgraph $\mathcal{B}$ closes as a loop at $\bar{\mathcal{I}}_1$, as required.

As a by-product of the above construction, there follows that the extreme left interval $\bar{\mathcal{I}}_{n-1}$ must ϕ-cover $\bar{\mathcal{I}}_{n-2}$, as well as all the intervals with odd indices (so that the piece-wise linear graph of f has a maximum at $\bar{\mathcal{I}}_{n-3} \cap \bar{\mathcal{I}}_{n-1} = p_{n-3}$), which completes the proof of the form of $\mathcal{S}$. If on the other hand, n is even and > 2 (to be more precise it is the smallest even period of periodic points of ϕ different from 2), the proof changes only at one point. The oddity of n was utilized only to deduce (9.103). We can therefore state that in the present case either the $\mathscr{A}$-graph of ϕ contains a subgraph $\mathscr{S}'$ which — by the same token as before — is of the form

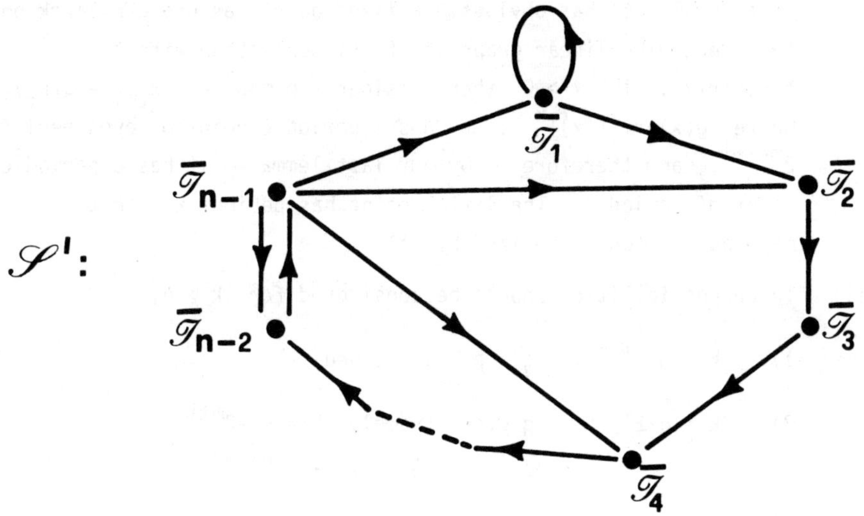

(notice that as $n-1$ is now odd, there are arrows from $\bar{\mathscr{I}}_{n-1}$ to all even vertices) or there is no element $\bar{\mathscr{I}}_{\bar{m}}$ of $\mathscr{A}$ for which (9.103) holds.

In the first case the loop $\bar{\mathscr{I}}_{n-1} \to \bar{\mathscr{I}}_{n-2} \to \bar{\mathscr{I}}_{n-1}$ gives a periodic point of period 2. In the second case the interval $\left[\min\limits_{1 \le i \le n} p_i, p_\ell\right]$, where p_ℓ denotes the left-end-point of $\mathscr{I}_1$, ϕ-covers the interval $\left[p_r, \max\limits_{1 \le i \le n} p_i\right]$, where p_r is the right-end-point of $\bar{\mathscr{I}}_1$, and vice versa. Thus again there exists a periodic point of period 2. In conclusion, if ϕ has a periodic point of even period $n > 2$, then it has a periodic point of period 2.

Suppose now that ϕ has a periodic point of period n, and consider the two possibilities:

i) $n = 2^m$, $m \ge 1$

ii) $n = p \cdot 2^m$, $m \geq 0$, $p > 1$, p odd .

i) $n \lhd k$ implies $k = 2^\ell$ with $\ell < m$. The case $\ell = 0$ $(k = 1)$ is trivial: ϕ has obviously a fixed point, as one can check on the piece-wise linear graph of f (intersection with the bisectrix). If $\ell > 0$ then consider the map ψ: $x_{t+1} = g(x_t)$, where $g(x) = [f(x)]^{k/2}$. ψ has a periodic point of even period $2^{m-\ell+1}$, and therefore — by our last lemma — it has a periodic point of period 2. The latter point has period k for the mapping ϕ (characterized by f).

ii) Three possibilities should be considered for $k \rhd n$,

1) $k = q \cdot 2^m$, $q > p$, q odd

2) $k = q \cdot 2^m$, q even (namely $k = q' \cdot 2^{m+t}$, $t > 0$, q' odd)

3) $k = 2^\ell$, $\ell \leq m$.

Consider the case 2) first. The map $\Theta = \phi^{2^m}$ has an $\mathscr{S}$-subgraph of the standard form, with $(p-1)$ vertices all representing intervals whose end-points have period p. Such a subgraph gives rise to the closed loops:

$$\mathscr{I}_1 \to \mathscr{I}_2 \to \cdots \to \mathscr{I}_{p-1} \to \mathscr{I}_1 \circlearrowleft (q\text{-}p) \text{ times} \quad , \quad \text{if } q > p$$

and

$$\mathscr{I}_{p-1} \to \mathscr{I}_{p-q} \to \cdots \to \mathscr{I}_{p-2} \to \mathscr{I}_{p-1} \quad , \quad \text{if } q < p \quad .$$

By (9.97), in both cases we obtain a periodic point of period q for Θ, which then gives ϕ-period k. Case 3) is easily proved by resorting to 2) in which one chooses $q = 1$, and then to i) (in which one replaces m with $m+1$). Case 1) can also be reconducted to 2). Indeed in this case $(q-p)$ is even and the same procedure utilized for case 2) leads to a claim whereby the periodic point of period q for Θ [which can now be ascribed to the existence of a loop of the first type above], has ϕ-period either of $q \cdot 2^m$ or $q \cdot 2^t$ for some $t < m$. In the first case the point has just ϕ-period k. In the second, one defines $n' = q \cdot 2^t$ $(n' \lhd n$ since $t < m)$ and applies then case 2) with $k = (q \cdot 2^{m-t})2^t$ (being $k \rhd n$, it is a fortiori $k \rhd n'$).

This completes the proof of Šarkovskiĭ's theorem. The consequences are far reaching; if there is a point $\bar{x}$ which has period 3, [i.e. $\bar{x}, f(\bar{x}), f^{(2)}(\bar{x})$ are all distinct and $f^{(3)}(\bar{x}) = \bar{x}$] then there are points $x_i \in \mathscr{I}$ of all possible, even arbitrarily large periods, even though an uncountable set $\mathscr{R} \subset \mathscr{I}$ exists such that no point $x \in \mathscr{R}$ is even asymptotically periodic. This situation is an example of what is referred to as <u>chaos</u>. More precisely we say that ϕ is chaotic if there exist a natural number t, and invariant set $\mathscr{K} \subset \mathscr{I}$ under ϕ^t and an equivalence relation $\sim$ on $\mathscr{K}$ such that if $\mu : \mathscr{K} \to \mathscr{K}/\sim$ is the canonical mapping induced by the equivalence relation ($\mathscr{K}/\sim$ is the space $\mathscr{K}$ in which all elements equivalent under $\sim$ are identified), then

$$\phi^{\#} = \mu\phi^t\mu^{-1} : \mathscr{K}/\sim \to \mathscr{K}/\sim \tag{9.107}$$

is conjugate to a (topological, symbolic) dynamics isomorphic to the Bernoulli shift defined in (7.26).

This gives a condition for the onset of chaos which is less restrictive than that stated above: indeed ϕ is chaotic if it has an r-periodic point with $r \neq 2^m$, for any $m \geq 0$. The latter statement follows from the fact that if ϕ is r-periodic, there are in all cases at least two subsets of $\mathscr{A}$ containing each element of $\mathscr{A}$ whose label has the same parity as r (see graphs $\mathscr{S}$ and $\mathscr{S}'$). Let $\mathscr{I}_0$ be one such element, and $\mathscr{I}_1 = \phi^m(\mathscr{I}_0) \neq \mathscr{I}_0$ for some suitable integer m. Then there are integers p and q such that

$$f^{(p)}(\mathscr{I}_0) \supset \mathscr{I}_0 \cup \mathscr{I}_1 \quad , \quad f^{(q)}(\mathscr{I}_1) \supset \mathscr{I}_0 \cup \mathscr{I}_1 \tag{9.108}$$

and the interval $\mathscr{K} \in \{f^{(k)}(\mathscr{J}_0)\ ,\ k = 1,\ldots,r-1\}$ such that $\mathscr{K} \supset \mathscr{J}_0 \cup \mathscr{J}_1$ is invariant under ϕ^{2pq}. The equivalence relation on $\mathscr{K}$ obtained by extending f linearly between adjacent orbit points makes ϕ^{2pq} conjugate to a baker's transformation.

We conclude by listing a set of mappings ϕ: $x_{n+1} = f(x_n)$, which have been solved exactly and which — depending on the choice of the initial condition — exhibit stable (convergent to a fixed point x^*), periodic or chaotic behaviour. The list is due to Katsura and Fukuda where the phase space is the plane (x_n, x_{n+1}); $q, p \in \mathbb{Z}, (q < p)$.

a.

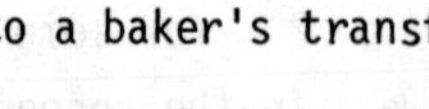

$$x_{n+1} = 4x_n(1 - x_n),\quad 0 \le x_n \le 1$$

$$x_n = \sin^2(2^n \sin^{-1}\sqrt{x_0})$$

$$x^* = 0\ ,\ 3/4$$

periodic if $x_0 = \sin^2\left(\pi \frac{q}{p}\right)$

b.

baker's transformation (see (7.4))

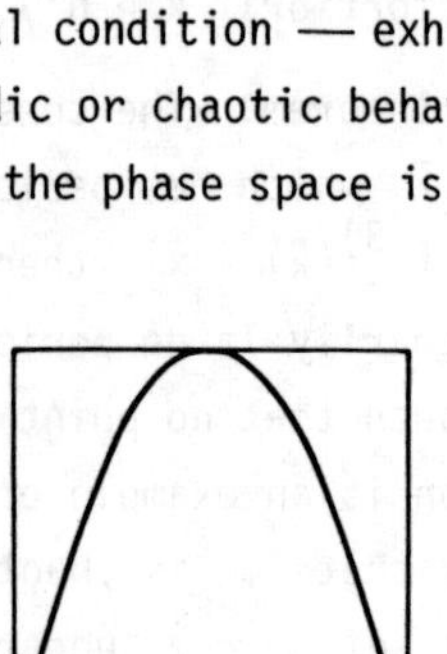

$$x_{n+1} = \begin{cases} 2x_n\ , & 0 \le x_n < \frac{1}{2} \\ 2(1 - x_n)\ , & \frac{1}{2} \le x_n \le 1 \end{cases}$$

$$x_n = \frac{1}{\pi}\cos^{-1}(\cos(2^n \pi x_0))$$

$$x^* = 0\ ,\ \frac{2}{3}$$

periodic if $x_0 = \frac{q}{p}$

c.

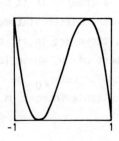

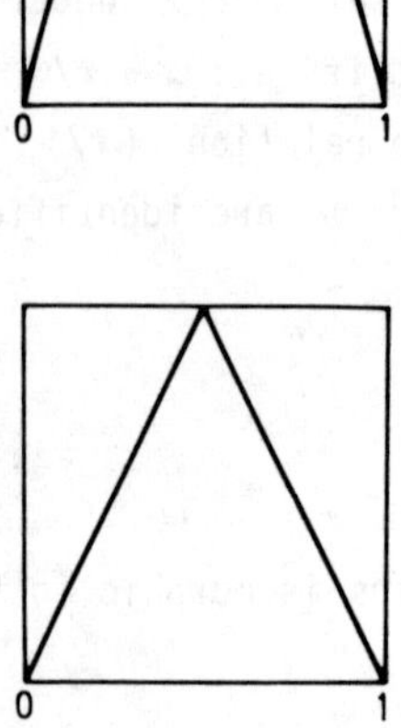

$$x_{n+1} = x_n(3 - 4x_n^2)\ ,\quad -1 \le x_n \le 1$$

$$x_n = \sin(3^n \sin^{-1} x_0)$$

$$x^* = 0\ ,\ \pm 1/\sqrt{2}$$

periodic if $x_0 = \sin\left(\pi \frac{q}{p}\right)$

d.

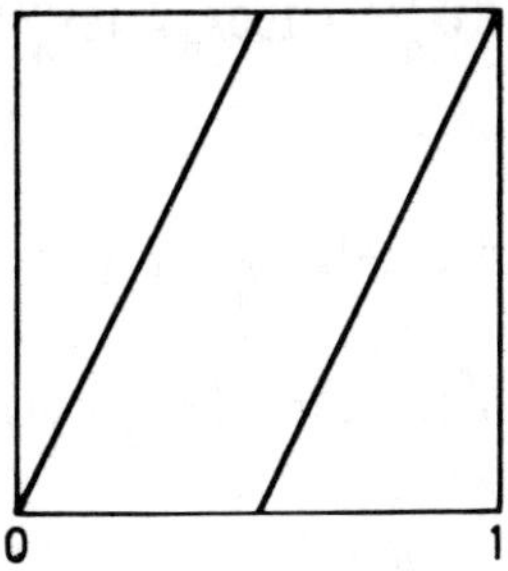

baker's transformation of the second kind (see (7.31)))

$$x_{n+1} = \begin{cases} 2x_n , & 0 \le x_n \le \frac{1}{2} \\ 2x_n - 1 , & \frac{1}{2} < x_n < 1 \end{cases}$$

$$x_n = \frac{1}{\pi} \cot^{-1}(\cot(2^n \pi x_0)) , \quad 0 \le x_n \le 1$$

$x^* = 0 , 1$

periodic if $x_0 = \frac{q}{p}$

e.

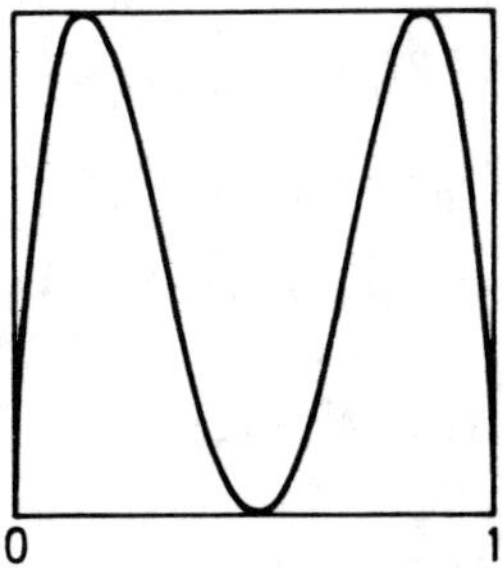

$$x_{n+1} = 16x_n(1 - x_n)(1 - 2x_n)^2 ,$$

$$0 \le x_n \le 1$$

$$x_n = \sin^2(4^n \sin^{-1} \sqrt{x_0})$$

$x^* = \frac{3}{4}$, $(5 \pm \sqrt{5})/8$

periodic if $x_0 = \sin^2 \left(\pi \frac{q}{p} \right)$

f.

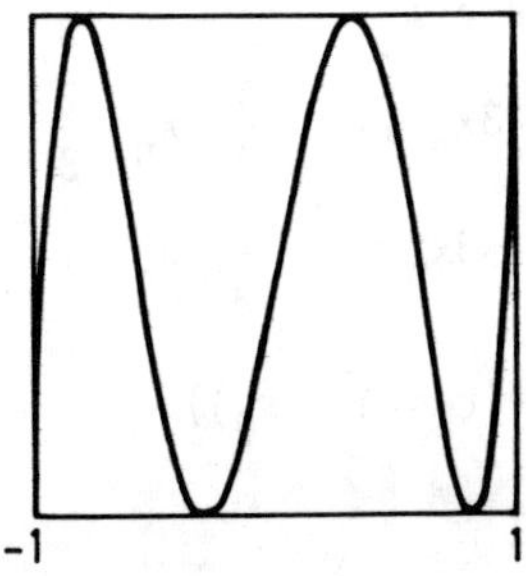

$$x_{n+1} = x_n(5 - 20x_n^2 + 16x_n^4) ,$$

$$-1 \le x_n \le 1$$

$$x_n = \sin(5^n \sin^{-1} x_0)$$

$x^* = 0 , \pm \frac{1}{2} , \pm 1$

periodic if $x_0 = \sin \left(\pi \frac{q}{p} \right)$

g.

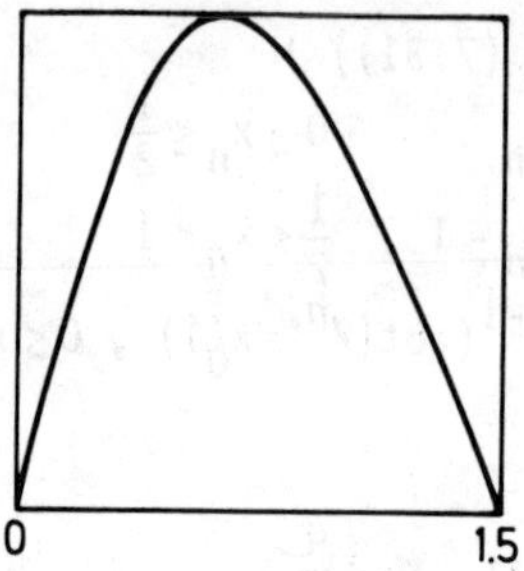

$$X_{n+1} = 24 - 84X_n^{1/2} + 100X_n - 48X_n^{3/2} + 8X_n^2$$

$$X_n = 1 + 2x_n$$

$$x_n = \sin^2(2^n \sin^{-1}(\sqrt{X_0} - 1)) + \frac{1}{2}\sin^4(2^n \sin^{-1}(\sqrt{X_0} - 1))$$

$$x^* = 0\ ,\ 1$$

periodic if $x_0 = \sin\left(\pi \frac{q}{p}\right) \times \left[1 + \frac{1}{2}\sin\left(\pi \frac{q}{p}\right)\right]$

h.

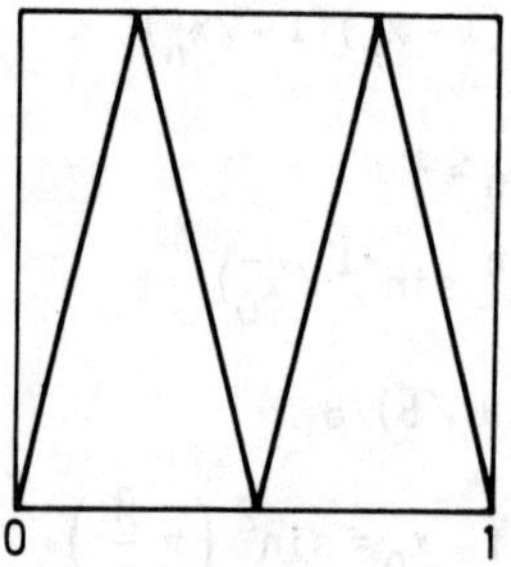

$$x_{n+1} = \begin{cases} 4x_n\ , & 0 \le x_n < \frac{1}{4} \\ 2(1 - 2x_n)\ , & \frac{1}{4} \le x_n < \frac{1}{2} \\ 2(2x_n - 1)\ , & \frac{1}{2} \le x_n < \frac{3}{4} \\ 4(1 - x_n)\ , & \frac{3}{4} \le x_n \le 1 \end{cases}$$

$$x_n = \frac{1}{\pi}\cos^{-1}(\cos(4^n\ \pi x_0))$$

$$x^* = 0\ ,\ \frac{2}{5}, \frac{2}{3}, \frac{4}{5}$$

periodic if $x_0 = \frac{q}{p}$

i.

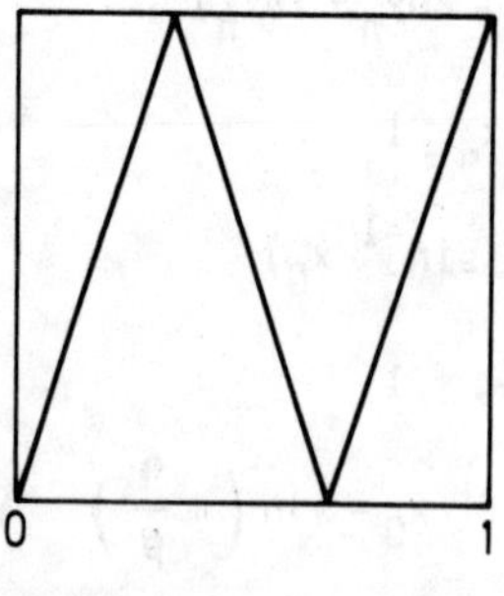

$$x_{n+1} = \begin{cases} 3x_n\ , & 0 \le x_n < \frac{1}{3} \\ 2 - 3x_n\ , & \frac{1}{3} \le x_n < \frac{2}{3} \\ -2 + 3x_n\ , & \frac{2}{3} \le x_n \le 1 \end{cases}$$

$$x_n = \frac{1}{\pi}\cos^{-1}(\cos(3^n\ \pi x_0))$$

$$x^* = 0\ ,\ \frac{1}{2}, 1$$

periodic if $x_0 = \frac{q}{p}$

j.

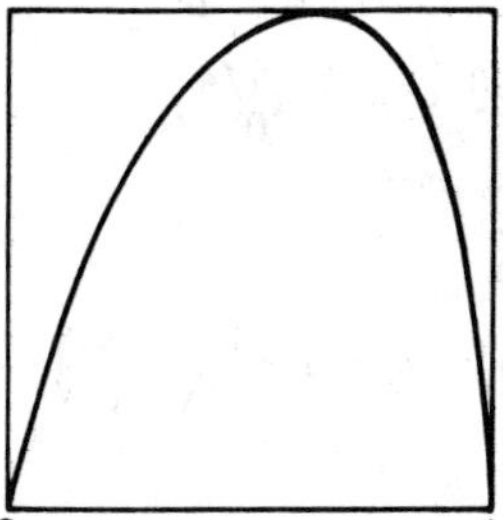

$$x_{n+1} = \frac{4x_n(1-x_n)(1-k^2x_n)}{(1-k^2x_n^2)^2} ,$$

$$0 \le x_n \le 1 , \quad 0 \le k^2 \le 1$$

$$x_n = \mathrm{sn}^2(2^n \,\mathrm{sn}^{-1}(\sqrt{x_0}, k), k)$$

$$x^* = 0 , \quad k^4x^{*4} - 3x^{*2} + 6x^* - 3 = 0$$

periodic if $x_0 = \mathrm{sn}^2\left(2\,\mathbb{K}(k)\,\frac{q}{p}, k\right)$

k.

$$x_{n+1} = \begin{cases} 2(x_n + \frac{1}{2}) , & -\frac{1}{2} \le x_n < -\frac{1}{4} \\ 2x_n , & -\frac{1}{4} \le x_n < \frac{1}{4} \\ 2(x_n - \frac{1}{2}) , & \frac{1}{4} \le x_n \le \frac{1}{2} \end{cases}$$

$$x_n = \frac{1}{\pi} \tan^{-1}(\tan(2^n \pi x_0))$$

$$x^* = 0$$

periodic if $x_0 = \frac{q}{p}$

l.

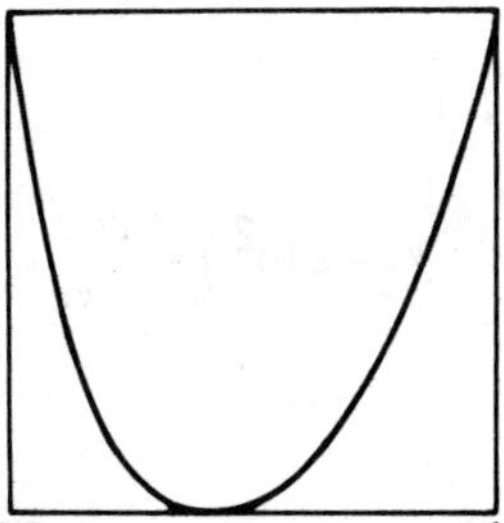

$$x_{n+1} = \left[\frac{1-k^2-2(1-k^2)x_n+x_n^2}{1-k^2-2x_n+x_n^2}\right]^2 ,$$

$$1-k^2 \le x_n \le 1$$

$$x_n = \mathrm{dn}^2(2^n \,\mathrm{dn}^{-1}(\sqrt{x_0}, k), k)$$

$x^* = 0.59479\ldots$ for $k^2 = \frac{1}{2}$

periodic if $x_0 = \mathrm{dn}^2\left(2\,\mathbb{K}(k)\,\frac{q}{p}, k\right)$

m.

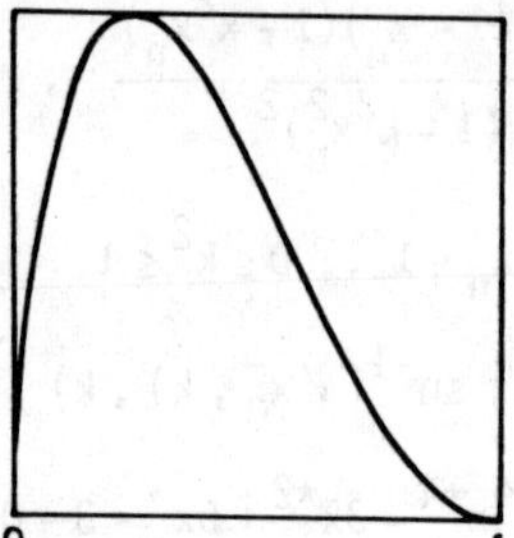

$$x_{n+1} = 16x_n(1 - 2\sqrt{x_n} + x_n)\ , \quad 0 \le x_n \le 1$$

$$x_n = \sin^4(2^n \sin^{-1} x_0^{1/4})$$

$$x^* = 0\ ,\ \frac{9}{16}$$

periodic if $x_0 = \sin^4\left(\pi\,\frac{q}{p}\right)$

n.

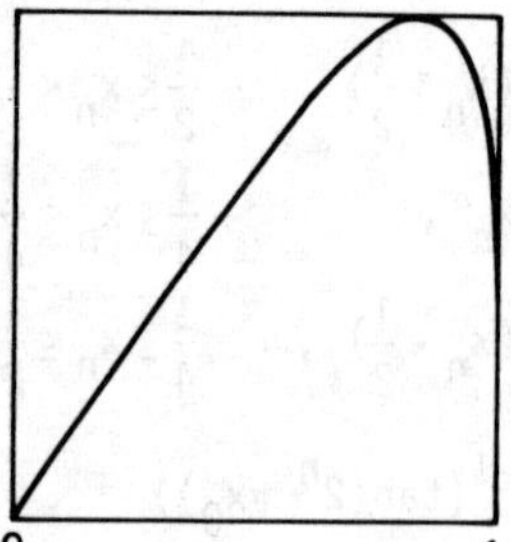

$$x_{n+1} = \sqrt{2}\,x_n(1 - x_n^4)^{1/4}\ , \quad 0 \le x_n \le 1$$

$$x_n = (\sin(2^n \sin^{-1} x_0^2))^{1/2}$$

$$x = 0\ ,\ \left(\frac{3}{4}\right)^{1/4}$$

periodic if $x_0 = \left[\sin\left(\pi\,\frac{q}{p}\right)\right]^{1/2}$

o.

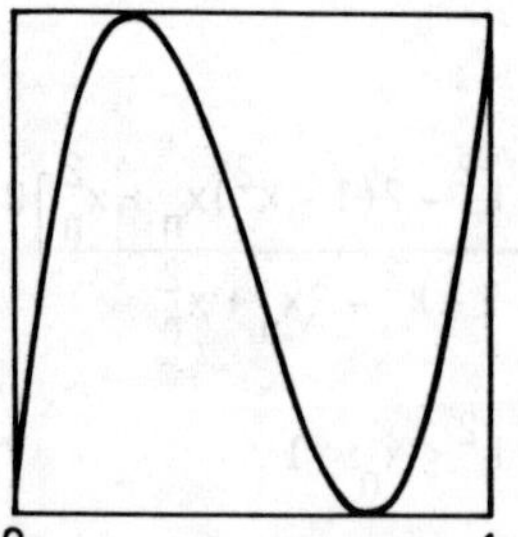

$$x_{n+1} = 16x_n^3 - 24x_n^2 + 9x_n\ , \quad 0 \le x_n \le 1$$

$$x_n = \frac{1}{2}\left[1 + \sin((-3)^n \sin^{-1}(2x_0 - 1))\right]$$

$$x^* = 0\ ,\ \frac{1}{2}\ ,\ 1$$

periodic if $x_0 = \sin^2\left(\pi\,\frac{2q+1}{4p}\right)$

p.

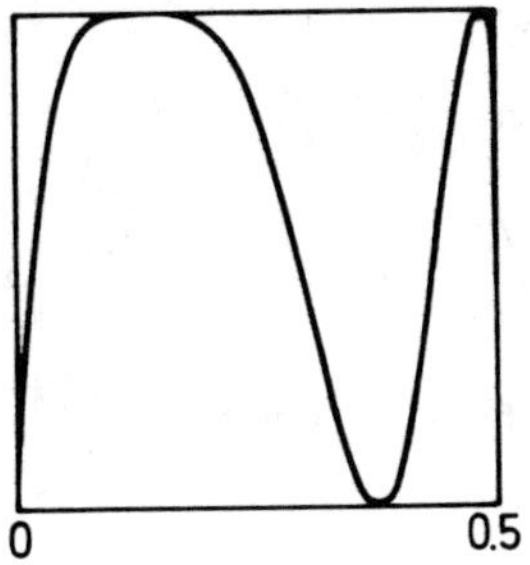

$$x_{n+1} = 16X_n(1 - 8X_n(1 - X_n)(1 - 2X_n)^2) \times (1 - X_n)(1 - 2X_n)^2$$

$$X_n = 1 - (1 - 2x_n)^{1/2} \, , \quad 0 \le x_n \le \frac{1}{2}$$

$$x_n = \sin^2(4^n \xi_0) - \frac{1}{2} \sin^4(4^n \xi_0)$$

$$\xi_0 = (\sin^{-1}(1 - (1 - 2x_0)^{1/2}))^{1/2}$$

$$x^* = 0 \, , \ (25 - 3\sqrt{5})/64 \, , \ \frac{15}{32} \, , \ 1$$

periodic if $x_0 = \sin\left[(\pi \frac{q}{p})^2\right] \times \left\{1 - \frac{1}{2} \sin\left[(\pi \frac{q}{p})^2\right]\right\}$

q.

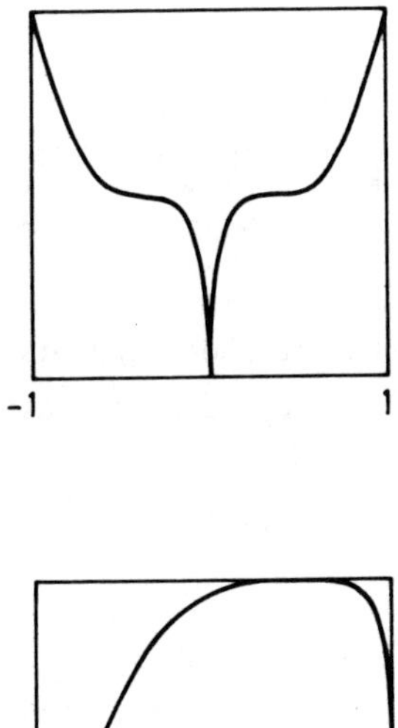

$$x_{n+1} = (2x_n^{2/3} - 1)^3 \, , \quad -1 \le x_n \le 1$$

$$x_n = \cos^3(2^n \cos^{-1} x_0^{1/3})$$

$$x^* = -\frac{1}{8}$$

periodic if $x_0 = \frac{1}{4}\left\{\cos\left(3\pi \frac{q}{p}\right) + 3 \cos\left(\pi \frac{q}{p}\right)\right\}$

r.

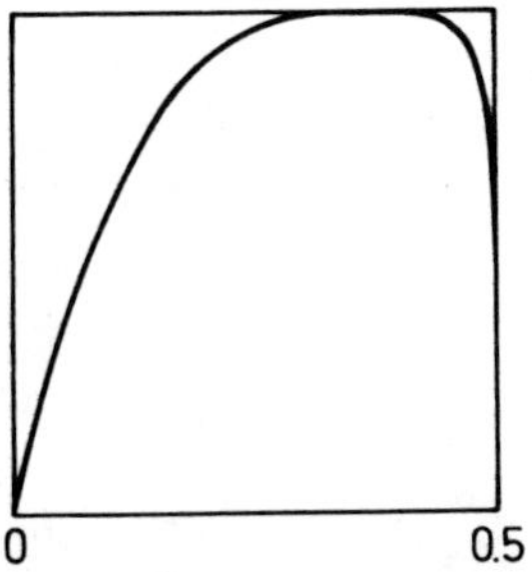

$$x_{n+1} = -X_n(8X_n^3 - 16X_n^2 + 12X_n - 4)$$

$$X_n = 1 - (1 - 2x_n)^{1/2} \, , \quad 0 \le x_n \le \frac{1}{2}$$

$$x_n = \sin^2(2^n \xi_0) - \frac{1}{2} \sin^4(2^n \xi_0)$$

$$\xi_0 = (\sin^{-1}(1 - (1 - 2x_0)^{1/2}))^{1/2}$$

$$x^* = \frac{15}{32}$$

periodic if $x_0 = \sin\left[\left(\pi \frac{q}{p}\right)^2\right] \times \left\{1 - \frac{1}{2} \sin\left[\pi\left(\frac{q}{p}\right)^2\right]\right\}$

s. 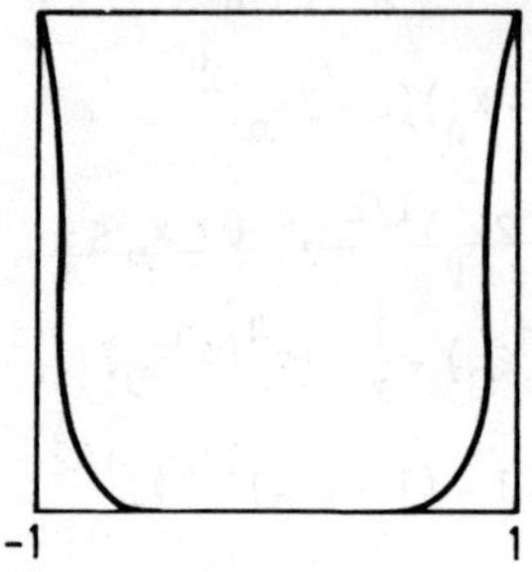

$$x_{n+1} = (2x_n^6 - 1)^{1/3}\,, \quad -1 \le x_n \le 1$$

$$x_n = (\cos(2^n \cos^{-1} x_0^3))^{1/3}$$

$$x = -\left(\frac{1}{2}\right)^{1/3}\,,\ 1$$

$$\text{periodic if} \quad x_0 = \left[\cos\left(\pi\,\frac{q}{p}\right)\right]^{1/3}$$

In the examples j and l, k is the modulus of the Jacobi elliptic function and $\mathbb{K}(k)$ the corresponding complete elliptic integral of the first kind

$$2\,\mathbb{K}(k) = \pi \prod_{n=1}^{\infty} (1 + k_n)\,,$$

$$k_n = \frac{1 - \sqrt{1 - k_{n-1}^2}}{1 + \sqrt{1 - k_{n-1}^2}}\,, \qquad k_0 \equiv k\,. \tag{9.109}$$

All the periodic solutions indicated become chaotic when q/p is replaced by an irrational number (π is replaced by $2\,\mathbb{K}$ in case of elliptic functions).

Chapter 2
Probability and Mechanics

1. The Gibbs Ensemble

The aim of statistical mechanics is to relate the observable properties of matter to the mechanical properties of the molecules constituting it. For systems of molecules the difficulty of predicting the possible outcome of experiment is twofold. On the one hand it stems from the fact that the number of particles can be so large that it is quite impossible to determine their dynamical state; on the other hand the dynamics can be so sensitive to initial conditions that their physical measurements (or experiments) are hard to reproduce. The initial state of the system is not controllable. In both cases the small scale fluctuations affect the system so much that only large scale behaviour can be predicted.

The mathematical technique to get rid of this kind of difficulty is the theory of probability. One attaches to each possible outcome of an experiment a suitably normalized number called its probability. If the non-reproducible experiment or the impractical measurements are repeated a very great number of times, then the number of times that each outcome effectively occurs should be proportional to the probability of that outcome. On the other hand, phenomenological observations give results which are indeed time averages, and the theory shall have to deal eventually with these in a self-consistent manner.

The way out of all these difficulties, due to the work of J. Willard Gibbs, is to represent the repetition of the experiments — instead of performed on the same system at different times — as being carried out at the same time on an enormous collection of different copies of the system. Such a collection is referred to as "Gibbs ensemble" and is characterized by systems (replicas of the real one) obeying identical dynamical laws and which are subjected to the same constraints and boundary conditions of the original system but with the molecules not necessarily in the same dynamical state. The ensemble is randomized in the sense that each configuration of velocity and position in the phase space of the real system accessible to the real system in its time evolution, is represented — at the same time — in the ensemble by one or more copies of the system itself; in such a way that the fraction of members of the ensemble whose phase space points lie within any given region of the phase space is equal to the probability that the outcome of the experiment should be found just in that region of phase space.

With this description in mind, it is quite natural to represent an ensemble in terms of a probability density function $\rho(P)$, where $P = \{q_\alpha, p_\alpha; \ \alpha = 1,\ldots,n\}$, $n = 3N$ is a generic point of phase space Φ, $P \in \Phi$; such that if $\mathscr{M}$ is any region of the phase space in which the system can be found,

$$\mathrm{Prob}(\mathscr{M}) \equiv \int_{\mathscr{M}} \rho(P)d\Omega \quad , \tag{1.1}$$

where $d\Omega$ denotes the Lebesgue volume element in Φ:

$$d\Omega = \prod_{\alpha=1}^{n} dq_\alpha dp_\alpha \quad . \tag{1.2}$$

Notice that $\rho(P) \geq 0$ and that, taking $\mathscr{M}$ to be the whole phase space, (1.1) gives the normalization condition

$$\int_{\Phi} \rho(P)d\Omega = 1 \quad . \tag{1.3}$$

It follows from the definition that if $F(P)$ is any dynamical variable, then its expected value — referred to as ensemble average — is given by

$$<F> = \int_{\Phi} F(P)\rho(P)d\Omega \qquad . \tag{1.4}$$

The fundamental problem of statistical mechanics is to determine what ensemble, i.e. which phase space probability density ρ, corresponds to any given physical situation. It is possible to state, under quite general conditions, three principles that the density ρ should satisfy: it turns out that such principles encompass — when combined with the dynamics of the physical system — all the information necessary to answer the fundamental question satisfactorily.

The first principle gives the relation between the phase space density at time t_0, $\rho_{t_0}(P)$ and at time t — this constitutes what is called the Liouville theorem. $\rho_{t_0}(P)d\Omega$ is the probability that at the initial time t_0 the system has to be found in a state represented by a point $P \in \Phi$ in the volume element $d\Omega$. In order to state the theorem, recall first that the canonical equations of motion, as discussed in Sec. I (the same symbols introduced there are used here), imply that there exists a transformation $G(t|t_0)$ on Φ, such that at time t the system will be found in the state represented by $P_t = \{q_\alpha(t), p_\alpha(t);\ \alpha = 1,\ldots,n\}$.

$$P_t = G(t|t_0)P_{t_0} \tag{1.5}$$

if it was at time t_0 in the state corresponding to P_0. We set

$$\dot{P} \equiv v(t, P) = \left\{ \frac{\partial H}{\partial q_\alpha}, -\frac{\partial H}{\partial p_\alpha} \quad ; \quad \alpha = 1,\ldots,n \right\} \tag{1.6}$$

where $H = H(P)$ is the Hamiltonian function, depending on the phase point P but — in general — not on the time t. It is worth recalling the group properties of G. To begin with,

$$G(t_0|t_0) = \mathbb{I} \tag{1.7}$$

is the identity transformation. Moreover, if the representative point gets to P_{t_2} at time t_2 after passing through P_1 at $t = t_1$, one can write

$$P_{t_2} = G(t_2|t_1)P_{t_1} = G(t_2|t_1)G(t_1|t_0)P_{t_0} \quad . \tag{1.8}$$

On the other hand, by definition (1.5),

$$P_{t_2} = G(t_2|t_0)P_{t_0} \quad . \tag{1.9}$$

Comparing (1.8) and (1.9) one gets the composition law

$$G(t_2|t_0) = G(t_2|t_1)G(t_1|t_0) \quad . \tag{1.10}$$

Moreover, since by the assumption of an Hamiltonian independent of time the motion is time reversible, for any t one can write, by (1.5) and (1.7)

$$P_{t_0} = G(t_0|t_0)P_{t_0} = G(t_0|t)G(t|t_0)P_{t_0} \tag{1.11}$$

whence one has the inverse element,

$$G(t_0|t) = [G(t|t_0)]^{-1} \quad . \tag{1.12}$$

Now, the dynamical variable $F = F(P)$ at time t will have the value $F_t = F(P_t)$, namely

$$F_t = F\left(G(t|t_0)P_{t_0} \right) \tag{1.13}$$

and its average will be given — according to (1.4) — by

$$<F_t> = \int_\Phi F_t(P)\rho(P)d\Omega =$$

$$= \int_\Phi F\left(G(t|t_0)P_{t_0}\right)\rho_{t_0}(P_{t_0})d\Omega_0 \quad . \tag{1.14}$$

Upon setting, for any $\bar{P}$ Φ, $\bar{P} = \{\bar{q}_\alpha, \bar{p}_\alpha; \quad \alpha = 1,\ldots,n\}$

$$\delta(\bar{P} - P) = \prod_{\alpha=1}^{n} \delta(\bar{q}_\alpha - q_\alpha)\delta(\bar{p}_\alpha - p_\alpha) \tag{1.15}$$

one has

$$F\left(G(t|t_0)P_{t_0}\right) = \int_\Phi \delta\left(G(t|t_0)P_{t_0} - P\right)F(P)d\Omega \tag{1.16}$$

and (1.14) can be written in the form

$$<F_t> = \int_\Phi F(P)\rho_{t,t_0}(P)d\Omega \tag{1.17}$$

where

$$\rho_{t,t_0}(P) = \int_\Phi \delta\left(G(t|t_0)P_{t_0} - P\right)\rho_{t_0}(P_{t_0})d\Omega_0 \tag{1.18}$$

is the probability density function at time t provided the probability density at time t_0 is $\rho_{t_0}(P)$. Notice that the definition (1.18) and (1.7) imply

$$\rho_{t_0,t_0}(P) = \int_\Phi \delta\left(P_{t_0} - P\right)\rho_{t_0}(P_{t_0})d\Omega_0 = \rho_{t_0}(P) \tag{1.19}$$

whence (1.17) reads also

$$<F_t> = \int_\Phi F\left(G(t|t_0)P\right)\rho_{t_0}(P)d\Omega \quad . \tag{1.20}$$

By the definitions (1.5) and (1.6)

$$\frac{\partial}{\partial t} G(t|t_0)P_{t_0} = v\left(t\,,G(t|t_0)P_{t_0}\right) \tag{1.21}$$

so that time differentiation of (1.18) gives

$$\frac{\partial}{\partial t}\rho_{t,t_0}(P) = \int_\Phi v\left(t\,,G(t|t_0)P_{t_0}\right)\delta'\left(G(t|t_0)P_{t_0} - P\right)\,\cdot$$
$$\cdot\,\rho_{t_0}(P_{t_0})d\Omega_0 \tag{1.22}$$

where δ' denotes the derivative of the Dirac delta function with respect to its argument. But

$$v\left(t\,,G(t|t_0)P_{t_0}\right)\delta'\left(G(t|t_0)P_{t_0} - P\right) =$$
$$= -\,\mathrm{grad}_P\left[v\left(t\,,G(t|t_0)P_{t_0}\right)\delta\left(G(t|t_0)P_{t_0} - P\right)\right]$$
$$= -\,\mathrm{grad}_P\left[v(t\,,P)\;\delta\left(G(t|t_0)P_{t_0} - P\right)\right] \tag{1.23}$$

hence

$$\frac{\partial}{\partial t}\rho_{t,t_0}(P) = -\,\mathrm{grad}_P\left[v(t\,,P)\int_\Phi \delta\left(G(t|t_0)P_{t_0} - P\right)\rho_{t_0}(P_{t_0})d\Omega_0\right]$$
$$= -\,\mathrm{grad}_P\left[v(t\,,P)\;\rho_{t,t_0}(P)\right]\quad. \tag{1.24}$$

In terms of canonical coordinates, (1.24) writes explicitly, using (1.6),

$$\frac{\partial}{\partial t}\rho_{t,t_0}(P) = -\sum_{\alpha=1}^{n}\left\{\frac{\partial}{\partial q_\alpha}\left[\frac{\partial H}{\partial p_\alpha}\rho_{t,t_0}(P)\right] - \frac{\partial}{\partial p_\alpha}\left[\frac{\partial H}{\partial q_\alpha}\rho_{t,t_0}(P)\right]\right\} =$$

$$= - \sum_{\alpha=1}^{n} \left(\frac{\partial H}{\partial p_\alpha} \frac{\partial}{\partial q_\alpha} \rho_{t,t_0}(P) - \frac{\partial H}{\partial q_\alpha} \frac{\partial}{\partial p_\alpha} \rho_{t,t_0}(P) \right)$$

$$= - \{\rho_{t,t_0}(P) , H\} \quad . \tag{1.25}$$

Equation (1.25) is the customary form of Liouville theorem, which can however be cast in a more useful form. By (1.18), if $\rho_{t_0}(P) = \text{constant}$, then $\rho_{t,t_0}(P) = \text{constant}$ as well at any t. In this case (1.17) and (1.20) imply

$$\int_\Phi F\Big(G(t|t_0)P\Big)\, d\Omega = \int_\Phi F(P) d\Omega \quad . \tag{1.26}$$

Let $\mathscr{A}$ be a finite subset of Φ, and choose for F in (1.26) the characteristic function of $\mathscr{A}$, i.e.

$$F(P) \equiv F_{\mathscr{A}}(P) = \begin{cases} 1 & \text{if} \quad P \in \mathscr{A} \\ 0 & \text{if} \quad P \notin \mathscr{A} \end{cases} , \tag{1.27}$$

then, due to (1.12)

$$F\Big(G(t|t_0)P \Big) = \begin{cases} 1 & \text{if} \quad P \in G(t_0|t)\mathscr{A} \\ 0 & \text{if} \quad P \notin G(t_0|t)\mathscr{A} \end{cases} \tag{1.28}$$

and (1.26) reads

$$\int_{\mathscr{A}} d\Omega = \int_{G(t|t_0)\mathscr{A}} d\Omega \quad . \tag{1.29}$$

$G(t_0|t)\mathscr{A}$ is the subset of Φ into which $\mathscr{A}$ evolves from time t_0 at time t. Then also,

$$\int_{\mathscr{A}} d\Omega = \int_{G(t_0|t)\mathscr{A}} d\Omega \quad . \tag{1.30}$$

The volume of the domain in Φ occupied at time t by the points which at time t_0 occupy $\mathscr{A}$ is equal to the volume of $\mathscr{A}$. In other words, Liouville's theorem states that the conservation of phase volume holds if the system evolution is described by the canonical equations of motion. On the other hand, choosing $\mathscr{A} \equiv \Phi$ (i.e. $F(P) = 1$), (1.17) and (1.20) give

$$\int_\Phi \rho_{t,t_0}(P)d\Omega = \int_\Phi \rho_{t_0}(P)d\Omega \quad . \tag{1.31}$$

By (1.18) and (1.31), we have therefore (recalling (1.3))

$$\rho_{t,t_0}(P) \geq 0 \quad , \quad \int_\Phi \rho_{t,t_0}(P)d\Omega = 1 \tag{1.32}$$

and $\rho_{t,t_0}(P)$ represents indeed — as expected — a probability distribution at time t.

Finally, select $F(P)$ of the form

$$F(P) = \mathscr{F}(P)\ \rho_{t_0}\Big(G(t_0|t)P\Big) \tag{1.33}$$

so that

$$F\Big(G(t|t_0)P\Big) = \mathscr{F}\Big(G(t|t_0)P\Big)\rho_{t_0}(P) \quad . \tag{1.34}$$

Upon inserting (1.34) into (1.26) one gets

$$\int_\Phi \mathscr{F}\Big(G(t|t_0)P\Big)\rho_{t_0}(P)d\Omega = \int_\Phi \mathscr{F}(P)\ \rho_{t_0}\Big(G(t_0|t)P\Big)d\Omega \quad . \tag{1.35}$$

On the other hand, once more equating the right-hand side of (1.17) and (1.20), one can write

$$\int_\Phi \mathscr{F}(P)\rho_{t_0}\Big(G(t_0|t)P\Big)d\Omega = \int_\Phi \mathscr{F}(P)\rho_{t,t_0}(P)d\Omega \quad . \tag{1.36}$$

Comparison of (1.35) and (1.36) — using the arbitrariness of the function $\mathscr{F}(P)$ — leads to the following statement of Liouville's theorem:

$$\rho_{t,t_0}(P) = \rho_{t_0}\left(G(t_0|t)P\right) \quad . \tag{1.37}$$

The second principle is usually taken for granted — or, more precisely, assumed as an implict axiom — yet it is extremely important in that it contains the notion of direction of time (i.e. the distinction between past and future). Referred to as the principle of causality, it states simply that the phase space density at any time is completely determined by what happened to the system before that time, and is unaffected by what will happen to the system in the future.

The last of the three principles, finally, is meant to rule out possibilities that a long-lasting phenomenological experience — or simply plain layman intuition — lead to consider unreasonable (for instance a gas in a container for which there were a non-vanishing probability for a set of exceptional configurations, such as the motion of the molecules is confined to a subset of the container). It requires that the probabilities characterizing the ensemble be really describable in terms of a phase space density $\rho(P)$, and that the latter were a well-behaved (smooth or at least piecewise continuous) function, rather than a general — possibly singular — measure.

Now, when a system is prepared in some way, and then left to itself, experience suggests that one can expect it to eventually settle down to some kind of equilibrium. Taking time t_0 as the moment when the preparation of the system finishes we can say, from the general principles of statistical mechanics stated before that;

i) due to the causality principle, the phase space density at time t_0 depends only on the preparation process (i.e. what happened to the system for $t < t_0$) and not on its evolution for $t > t_0$;

ii) the phase space density ρ_{t,t_0} at any $t > t_0$ can be calculated, using Liouville theorem, from ρ_{t_0}

$$\rho_{t,t_0}(P) = \rho_{t_0}\left([G(t|t_0)]^{-1}P \right) \quad . \tag{1.37'}$$

iii) the expected value of any dynamical variable F at time $t > t_0$, defined by (1.17), can finally be calculated, due again to the Liouville's theorem, by (1.20), and depends indeed once more only on ρ_{t_0}:

$$< F_t > = < F >_t \quad . \tag{1.38}$$

In fact, one knows very little about ρ_{t_0}; yet equilibrium statistical machanics can be developed from the three principles stated, in such a way that it gives a great deal of information about the equilibrium value

$$F_{eq} = \lim_{t \to \infty} < F >_t \quad . \tag{1.39}$$

The problem of computing F_{eq}, when it does exist, is in the realm of ergodic theory. The latter is a rapidly growing mathematical subject, which has recently produced a wealth of results of great interest in relation to statistical mechanics. Here only the basic relevant elements shall be reviewed (a more formal approach to the concepts used here, will be found in the last section of the book).

To begin with, one should notice that F_{eq}, as defined in (1.39) does not always exist, in that $< F >_t$ may well be a quantity forever oscillating rather than approaching a limit. On the other hand, if it exists, it is equal to

$$F_{eq} = \lim_{T \to \infty} \frac{1}{(T - t_0)} \int_{t_0}^{T} < F >_t \, dt \tag{1.40}$$

which always exists.

In order to avoid trying to calculate a quantity which may not exist, (1.40) should be assumed to replace (1.39) as definition of F_{eq}. From (1.40) there follows, setting the time origin so that $t_0 = 0$, and writing,

$$F_{eq} = \lim_{T\to\infty} \frac{1}{T}\int_0^T dt \int_\Phi F(P)\rho_{t,0}(P)d\Omega \equiv$$

$$\equiv \int_\Phi F(P)\tilde{\rho}(P)d\Omega \tag{1.41}$$

that the definition of the time-averaged phase space density is

$$\tilde{\rho}(P) = \lim_{T\to\infty} \frac{1}{T}\int_0^T \rho_{t,0}(P)dt =$$

$$= \lim_{T\to\infty} \frac{1}{T}\int_0^T \rho_0\left([G(t|0)]^{-1}P \right) dt \qquad . \tag{1.42}$$

The interchange of the operations of limit and integration in the definition (1.41) is justified by Birkhoff's theorem, stating that $\tilde{\rho}(P)$ is well-defined at almost all phase points (namely the set of points $P' \in \Phi$ where (1.41) is violated has measure zero in Φ), provided $F(P)$ has the same regularity properties as $\rho_0(P)$, i.e. is piece-wise continuous. Once the existence of the limit in (1.42) is granted, it is an immediate consequence of the group property of the transformation G, (1.10), that

$$\tilde{\rho}(P) = \tilde{\rho}\left(G(t|0)P \right) \quad , \quad \forall t \tag{1.43}$$

i.e. $\tilde{\rho}$ is constant on the trajectories in phase space. A probability density satisfying (1.43) is said to be invariant, because — as required by Gibbs' definition — the properties of the ensemble characterized by such density do not change with time.

For conservative systems, the Hamiltonian too is constant along every trajectory, thus a way of satisfying Eq. (1.43) is for $\tilde{\rho}$ to be a function of the Hamiltonian,

$$\tilde{\rho}(P) = \psi(H(P)) \quad . \tag{1.44}$$

If it happens that all the invariant probability densities have the form (1.44), the system is said to be ergodic. The interest of ergodic systems is that the calculations of averages are replaced by integrals over the energy manifold $\mathscr{E}$, defined by an equation of the form $H(P) = E = \text{const.}$, for some fixed energy value E.

Noticing that a measure $\mu_E(P)$, $P \in \mathscr{E}$ on $\mathscr{E}$ can be defined by

$$\int_{\mathscr{E}} f(P)d\mu_E(P) = \int_{\Phi} f(P)\delta(H(P) - E)d\Omega \tag{1.45}$$

where f is any continuous function on Φ, upon setting

$$\Omega_E = \text{vol}(\mathscr{E}) = \int_{\mathscr{E}} d\mu_E(P) = \int_{\Phi} \delta(H(P) - E)d\Omega \tag{1.46}$$

averages over $\mathscr{E}$ can be defined as

$$< F >_E = \frac{1}{\Omega_E} \int_{\mathscr{E}} F(P)d\mu_E(P) \quad . \tag{1.47}$$

$< F >_E$ is called the microcanonical average of $F(P)$ and can be thought of as the average of $F(P)$ in an ensemble (the "microcanonical ensemble") whose systems have all the same energy E.

From (1.41), using (1.44) and (1.47) one gets

$$F_{eq} = \int_{\Phi} F(P)\tilde{\rho}(P)d\Omega = \int_{\Phi} F(P)\psi(H(P))d\Omega = \int_{\Phi} F(P)\left\{ \int_{-\infty}^{+\infty} \psi(E)\delta(E - H(P))dE \right\} d\Omega =$$

$$= \int_{-\infty}^{+\infty} \psi(E) \left\{ \int_{\Phi} F(P)\delta(E - H(P))d\Omega \right\} dE = \int_{-\infty}^{+\infty} \psi(E) \left\{ \int_{\mathscr{E}} F(P)d\mu_E(P) \right\} dE \equiv$$

$$\equiv \int_{-\infty}^{+\infty} p(E) < F >_E dE \tag{1.48}$$

which defines the probability density

$$p(E) = \Omega_E \psi(E) \quad . \tag{1.49}$$

By (1.46), (1.44)

$$p(E) = \psi(E)\int_\Phi \delta(E - H(P))d\Omega = \int_\Phi \psi(H(P))\delta(E - H(P))d\Omega =$$

$$= \int_\Phi \tilde{\rho}(P)\delta(E - H(P))d\Omega \quad . \tag{1.50}$$

But, since $H(P)$ is a constant of the motion, recalling (1.31) (with $t_0 = 0$) and (1.41) one can write

$$p(E) = \int_\Phi \rho_0(P)\delta(E - H(P))d\Omega =$$

$$= \int_{\mathscr{E}} \rho_0(P)d\mu_E(P) = \Omega_E < \rho_0 >_E \quad . \tag{1.51}$$

Now (1.48) can be interpreted as stating that if the system is ergodic, one can think of the equilibrium ensemble with density $\tilde{\rho}(P)$ as a mixture of microcanonical ensembles; their proportions in the mixture being given by the probability density $p(E)$ of the dynamical variable H in the time averaged ensemble as defined by (1.51). This makes the microcanonical ensemble the only one effectively relevant in constructing statistical mechanics. Notice also that comparison between (1.51) and (1.49) yields

$$\psi(E) = < \rho_0 >_E \quad . \tag{1.52}$$

Moreover, ergodicity is by no means a universal property of mechanical systems (indeed some of the simplest and most important systems used in statistical mechanics, such as the ideal gas or the multiple harmonic oscillator, are not ergodic).

In the microcanonical ensemble, the simplest probability density, obviously satisfying (1.43) is $\rho_0(P) = \text{const} \cdot \delta(H(P) - E)$. Since in this case one wants $F_{eq} \equiv \langle F \rangle_E$, namely $p(E') = \delta(E' - E)$, the constant — due to the normalization condition (1.32) as well as to (1.49) — should be Ω_E^{-1}, namely

$$\rho_0(P) = \begin{cases} \dfrac{1}{\Omega_E} & \forall P \in \mathscr{E} \\ 0 & \text{elsewhere} \quad (\Phi \setminus \mathscr{E}) \end{cases} \quad . \tag{1.53}$$

Of course then, $\rho_{t,0}(P) = 1/\Omega_0 = \tilde{\rho}(P)$. It is interesting to notice that it is also unique, i.e. it is the only one invariant. In order to prove this fundamental feature, let us notice first that (1.53) implies a much stricter definition of ergodicity. Indeed, one has in this case — resorting to (1.26) — that

$$\langle F \rangle_E = \lim_{T \to \infty} \frac{1}{T} \int_0^T F\Big(G(t|0)P\Big)\, dt \quad . \tag{1.54}$$

In other words, the time average of the dynamical variable F is equal to its average over the microcanonical ensemble (for almost all $P \in \mathscr{E}$). Equation (1.54) is noteworthy in that its right-hand side — denote it $\bar{F}(P)$ — is in principle a function of P, whereas (1.54) requires it to be independent on P, i.e. constant over the whole $\mathscr{E}$ (except possibly for a set of measure μ_E zero).

Let us now move to the proof of the uniqueness of $\rho_0(P)$ as given in (1.53). We observe first that if a dynamical system is such that its energy manifold $\mathscr{E}$ is the disjoint union of two sub-manifolds $\mathscr{E}_1, \mathscr{E}_2$ (each of non-zero measure), each of which is invariant under $G(t|0)$,

$$G(t|0)\mathscr{E}_1 = \mathscr{E}_1 \quad , \quad G(t|0)\mathscr{E}_2 = \mathscr{E}_2 \quad \forall t > 0 \tag{1.55}$$

then such a system — which is called decomposable — is not ergodic. This is easily done considering the special case in which the dynamical variable F is the characteristic function (in the sense of (1.27)) of one of the sub-manifolds, say $\mathscr{E}_1$

$$F_{\mathscr{R}}(P) = \begin{cases} 1 & \text{if} \quad P \in \mathscr{E}_1 \quad (P \notin \mathscr{E}_2) \\ 0 & \text{if} \quad P \notin \mathscr{E}_1 \quad (P \in \mathscr{E}_2) \end{cases} \tag{1.56}$$

and realizing that $\bar{F}_{\mathscr{R}}(P)$ in this case depends on P, and hence cannot be equal to $\langle F \rangle_E$. Also the inverse statement holds: a non-ergodic system is decomposable. In fact, the property of non-ergodicity implies the existence of a dynamical function $F'(P)$ whose time average $\bar{F}'(P)$ is <u>not</u> constant almost everywhere (with no loss of generality one can assume such a function to be real). But then, upon setting

$$\bar{\mathscr{E}}_1 = \{P | \bar{F}'(P) < \mathscr{F}\} \quad \text{and}$$

$$\bar{\mathscr{E}}_2 = \{P | \bar{F}'(P) \geq \mathscr{F}\} \tag{1.57}$$

where $\mathscr{F}$ is a real constant, which can be chosen so that both $\bar{\mathscr{E}}_1$ and $\bar{\mathscr{E}}_2$ have positive measures, we notice that — due to the property that, by its very definition, time average is invariant with respect to $G(t|0)$ — $\bar{\mathscr{E}}_1$ and $\bar{\mathscr{E}}_2$ are decomposable.

Thus, a dynamical system is <u>ergodic</u> if and only if it is not decomposable, namely <u>if every invariant set has measure 1</u>, and then <u>if every invariant function is constant almost everywhere</u>. Let us suppose that the system is prepared in such a way as to be represented at $t = 0$ by the point $P \in \mathscr{E}$, and that $F(P)$ is quite far from F_{eq}. As time passes, one expects the value of F, $F(G(t|0)P)$, to become closer and closer to F_{eq}, so that as $t \to \infty$ the system spends most of the time in proximity of such equilibrium state — from which only occasionally it should depart, due to rare, large random fluctuations. Since both the initial transient and the fluctuations should contribute little or nothing to $\bar{F}(P)$, by its definition the ergodic theorem proved before, states that a microcanonical ensemble, for which $\bar{F}(P)$ is almost always equal to $\langle F \rangle_E$, is well designed to describe situations in which the the fluctuations around the equilibrium configuration are controllable.

An equivalent way of stating the feature of ergodicity is the following. For a given observable event A, let $\mathscr{R} \subset \mathscr{E}$ be the region in the manifold $\mathscr{E}$ compatible with A (i.e. such that the event A is observed at time t if and only if the phase point of the system at time t is in $\mathscr{R}$). If the system is observed for a very long time, the fraction of time in which the event A is observed is obviously given by the time average $\bar{F}_{\mathscr{R}}(P)$ of the characteristic function $F_{\mathscr{R}}$ of $\mathscr{R}$, where P is e.g. the initial dynamical state of the system. By the ergodic theorem, such quantity is equal to the ensemble average,

$$\bar{F}_{\mathscr{R}}(P) = \langle F_{\mathscr{R}} \rangle_E = \frac{\int_{\mathscr{R}} d\mu_E(P)}{\int_{\mathscr{E}} d\mu_E(P)} \quad . \tag{1.58}$$

The right-hand side of (1.58) is nothing but (recall definition (1.1)) the probability — defined in the microcanonical ensemble — that the event A happens. Thus for an ergodic system, the probability of an event equals the fraction of time in which it is observed.

Even when the system is ergodic, so that (1.48) can be used for all equilibrium values by resorting to (1.54), there remains the question whether the expectations of dynamical variables actually approach these equilibrium values as the system evolves. It was already pointed out that one should replace (1.39) by (1.40) to get rid of those cases in which the limit does not exist ($\langle F \rangle_t$ oscillating). But even when it does, the question

$$\lim_{t \to \infty} \int_{\Phi} F(P)\rho_{t,0}(P)d\Omega \overset{?}{=} F_{eq} \tag{1.59}$$

remains open. In order to deal more properly with it, let us rewrite the integral at the left-hand side of (1.59) using the same expansion which lead to (1.48):

$$\int_{\Phi} \rho_{t,0}(P)F(P)d\Omega = \int_{-\infty}^{+\infty} dE \; \Omega_E < \rho_{t,0}(P)F(P) >_E \tag{1.60}$$

(where the definition (1.46) was used). Using (1.60) on the left side of (1.59), and (1.48) — in which (1.49) and (1.52) are utilized to express $p(E)$ explicitly — on the right, (1.59) can be rephrased as whether (assume the limit exists)

$$\lim_{t\to\infty} \int_{-\infty}^{+\infty} dE\ \Omega_E < \rho_{t,0}(P)F(P) >_E \stackrel{?}{=} \int_{-\infty}^{+\infty} dE\ \Omega_E < \rho_0 >_E < F >_E \ . \qquad (1.61)$$

A <u>sufficient</u> (by no means necessary) condition for (1.61) to be answered affirmatively is that, for every E, and for every pair of regular functions $\rho_0(P)$, $F(P)$ on $\mathscr{E}$ (square integrable with respect to the microcanonical measure μ_E)

$$\lim_{t\to\infty} < \rho_0\Big(G(0|t)P \Big) F(P) >_E = < \rho_0 >_E < F >_E \qquad (1.62)$$

holds ((1.37') and (1.12) were used). In that case the system is said to be <u>mixing</u>. It is obvious that mixing implies ergodicity — as one can straightforwardly show replacing in (1.62) P by $G(t|0)P$ and using (1.12), whereby (1.62) reduces to (1.54) —, whereas in general ergodicity does not imply <u>mixing</u>.

The physical meaning of mixing is the following. Let $\mathscr{A}$ and $\mathscr{B}$ be subsets of $\mathscr{E}$, and let the representative points of the system at time 0 be in the set $\mathscr{B}$ (i.e. $\rho_0(P) = F_{\mathscr{B}}(P)$). If the system is mixing, one can check from (1.62), by choosing F to be the characteristic function of $\mathscr{A}$, $F_{\mathscr{A}}(P)$, that the fraction of members of the ensemble which are in the set $\mathscr{A}$ at time t, for $t\to\infty$ tends to a limit which is just the fraction of the area of $\mathscr{E}$ occupied by $\mathscr{A}$. In other words the points even if originally localized in a set $\mathscr{B}$ distribute themselves evenly over the entire space with the passage of time and it is difficult to distinguish points which were or were not in some other set a sufficiently long time earlier.

Moreover, let $\rho_0(P)$ be a non-negative measurable function, such that

$$\int_{\mathscr{E}} \rho_0(P)d\mu_E(P) = 1 \quad . \tag{1.63}$$

Introduce a new measure ν_0, absolutely continuous with respect to μ_E, whose Radon-Nikodym derivative equals ρ_0:

$$\frac{d\nu_0(P)}{d\mu_E(P)} = \rho_0(P) \quad . \tag{1.64}$$

ν_0 is in general different from μ_E, thus it represents a non-equilibrium measure. One can then determine the time evolution of ν_0, setting

$$\nu_t(P) = \nu_0\left(\left[G(t|0)\right]^{-1}P\right) \quad . \tag{1.65}$$

Then in the case of a transformation with mixing, by (1.62)

$$\int_{\mathscr{E}} F(P)d\nu_t(P) = \int_{\mathscr{E}} F(P)\rho_0\left(G(t|0)P\right)d\mu_E(P) \xrightarrow[t\to\infty]{} \int_{\mathscr{E}} F(P)d\mu_E(P) \quad . \tag{1.66}$$

Thus the non-equilibrium distribution tends toward equilibrium.

2. Instability

It is in general quite difficult to prove whether a given mechanical system has the properties of ergodicity and proofs have been given so far only for a few models describing physical systems (the most important being certain hard sphere systems enclosed in a box with perfectly reflecting walls, thoroughly studied by Sinai who was able to prove them both ergodic and mixing) and abstract dynamical flows. The essential feature of these flows is a particular type of instability of their orbits, whereby in general two points initially very close in phase space move apart with exponential rapidity as the flow proceeds. In order to discuss this property, let us consider a simple example, constructed after Smale and Arnold.

To begin with, consider a toroidal phase space:

$$\Phi \equiv \{(q, p) \bmod 1\} \tag{2.1}$$

with the usual measure $d\mu = dqdp$ and the mapping, describing an automorphism of Φ,

$$\tilde{\mathcal{T}} : (q, p) \rightarrow (q+p, q+2p) (\bmod 1) \quad . \tag{2.2}$$

Such a mapping induces on the covering space of Φ (the plane (q, p)) a mapping $\mathcal{T}$, which — being linear — is represented by the matrix,

$$\mathcal{T} = \begin{pmatrix} 1 & 1 \\ 1 & 2 \end{pmatrix} \quad . \tag{2.3}$$

Since $\det \mathcal{T} = 1$, $\tilde{\mathcal{T}}$ is measure preserving. $\mathcal{T}$ has two real eigenvalues $\tau_1, \tau_2 : 0 < \tau_2 < 1 < \tau_1$ $(\tau_{1,2} = \frac{1}{2}\{3 \pm \sqrt{5}\})$. Given a set $A \subset I \times I$ (where I is the interval $0 — 1$), $\mathcal{T}^n \circ A$, for n large approaches a long, thin band in the plane (q, p), in the neighbourhood of an orbit of the dynamical system

$$\dot{q} = 1 \quad , \quad \dot{p} = \tau_1 - 1 \quad . \tag{2.4}$$

This can be easily checked in the following manner. Compute first $\mathcal{T}^n$. Upon setting

$$\mathcal{T}^n \equiv \begin{pmatrix} a_n & b_n \\ c_n & d_n \end{pmatrix} \tag{2.5}$$

one easily finds (writing $\mathcal{T}^{n+1} = \mathcal{T}^n \cdot \mathcal{T}$, with $\mathcal{T}$ given by (2.3)) that the matrix elements a_n, b_n, c_n, d_n are given by the following relations

$$b_n = c_n = d_n - a_n$$

$$\begin{cases} a_{n+1} = d_n \\ d_{n+1} = 3d_n - a_n \quad , \quad n \geq 1 \end{cases} \quad . \tag{2.6}$$

The latter recursion reads, after eliminating d_n,

$$a_{n+2} = 3a_{n+1} - a_n \quad , \quad a_1 = 1 \quad , \quad a_2 = 2 \quad . \tag{2.7}$$

Resorting to the generating function

$$\phi(x) \equiv \sum_{n=1}^{\infty} a_n x^n \quad , \quad x \in \mathbb{R} - (\tau_1 , \tau_2) \tag{2.8}$$

it is straightforward to verify that (2.7) implies

$$\phi(x) = \frac{x(1-x)}{(\tau_1 - x)(\tau_2 - x)} \tag{2.9}$$

whence, after some simple algebra,

$$a_n = \frac{1-\tau_2}{\tau_1 - \tau_2} \tau_1^n \left[1 + \tau_1 \left(\frac{\tau_2}{\tau_1}\right)^n\right] \quad . \tag{2.10}$$

and, by (10.6),

$$b_n = c_n = \frac{1}{\tau_1 - \tau_2} \tau_1^n \left[1 - \left(\frac{\tau_2}{\tau_1}\right)^n\right] \tag{2.11}$$

$$d_n = \frac{\tau_1 - 1}{\tau_1 - \tau_2} \tau_1^n \left[1 + \tau_2 \left(\frac{\tau_2}{\tau_1}\right)^n\right] \quad . \tag{2.12}$$

Notice that $a_n d_n - b_n c_n = \det(\mathcal{T}^n) = 1$, as expected.

Assuming now $\mathcal{A}$ to be a square (whose sides, parallel to the coordinate axes have length ℓ), it is mapped under $\mathcal{T}^n$ into a parallelogram $\mathcal{A}'$ (see figure next page):

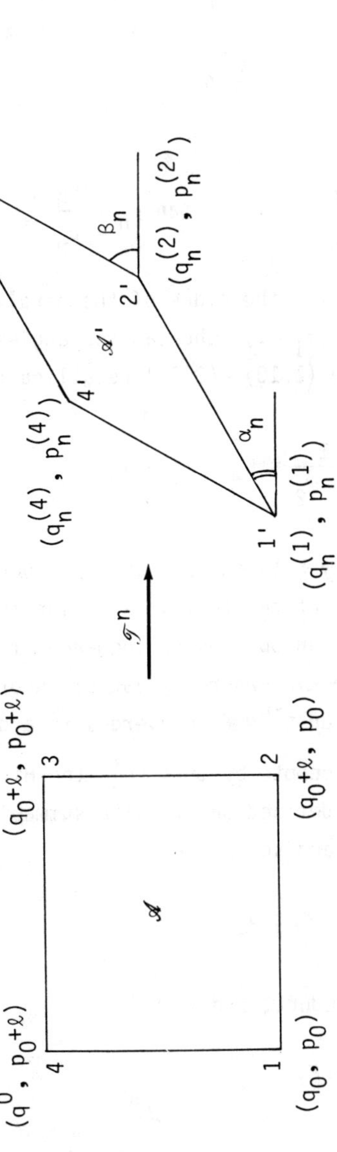

(q_0, p_0) 1
2 $(q_0+\ell, p_0)$
3 $(q_0+\ell, p_0+\ell)$
4 $(q^0, p_0+\ell)$
$\mathcal{A}$
$\mathcal{T}^n$
1' $(q_n^{(1)}, p_n^{(1)})$
2' $(q_n^{(2)}, p_n^{(2)})$
3' $(q_n^{(3)}, p_n^{(3)})$
4' $(q_n^{(4)}, p_n^{(4)})$
$\mathcal{A}'$
α_n
β_n

The vectors $(1' \to 2')$, $(2' \to 3')$ are given, respectively, by:

$$\ell \begin{pmatrix} a_n \\ c_n \end{pmatrix} \quad , \quad \ell \begin{pmatrix} b_n \\ d_n \end{pmatrix}$$

whence

$$\tan \alpha_n = \frac{c_n}{a_n} \quad , \quad \tan \beta_n = \frac{d_n}{b_n} \quad . \tag{2.13}$$

In the limit $n \to \infty$, the sides of the parallelogram $\mathscr{A}'$ become infinitely long $(\tau_1 > 1)$ whereas the angles α_n, β_n tend to the common value (use (2.10) - (2.12) recalling that $\tau_2/\tau_1 < 1$)

$$\theta \equiv \alpha_\infty = \frac{1}{1 - \tau_2} = \beta_\infty = \tau_1 - 1 \tag{2.14}$$

where the equality between α_∞ and β_∞ depends on the fact that τ_1, τ_2 are solutions of the secular equation for $\mathscr{T}$, $\tau^2 - 3\tau + 1 = 0$. Equation (2.4) is an obvious consequence of (2.14). On Φ, the orbit of (2.4) is dense everywhere by Jacobi theorem, since $(\tau_1 - 1)$ is irrational and then $\tilde{\mathscr{T}}^n \circ \mathscr{A}$ converges to a dense helix on the torus.

Let us now denote by $\mathscr{H}(\Phi, \mu)$ the Hilbert space of the complex valued functions defined on Φ with summable square with respect to the measure $d\mu$ and set

$$< f | g > = \int_\Phi f \cdot \bar{g} \, d\mu \tag{2.15}$$

for the scalar product, and

$$\| f \| = \sqrt{< f | f >} \quad , \tag{2.16}$$

for the norm in $\mathscr{H}$. The mapping $\tilde{\mathscr{T}}$ (in general any mapping defining an (abstract) dynamical system on a phase manifold Φ) induces an operator $\mathscr{U}$ on $\mathscr{H}$, defined by

$$\mathcal{U}f(P) = f(\tilde{\mathcal{T}}(P)) \quad , \quad \forall\ f \in \mathcal{H} , \quad P \in \Phi \quad . \tag{2.17}$$

$\mathcal{U}$ is obviously linear and bijective $(\mathcal{U}f \in \mathcal{H})$, moreover it is isometric if $\tilde{\mathcal{T}}$ is measure preserving:

$$\|\mathcal{U}f\|^2 = \int_\Phi |f(\tilde{\mathcal{T}}(P))|^2 \, d\mu(P) = \int_\Phi |f(\tilde{\mathcal{T}}(P))|^2 \, d\mu(\tilde{\mathcal{T}}(P)) =$$

$$= \int_\Phi |f(P)|^2 d\mu(P) = \| f \|^2 \quad . \tag{2.18}$$

Thus $\mathcal{U}$ is a unitary operator of $\mathcal{H}$, and its eigenvalues can be written as $\{\omega = e^{i\tau}, \ \tau \in \mathbb{R}\}$ i.e. they have absolute value 1. Let f be the eigenfunction corresponding to ω, namely

$$f(\tilde{\mathcal{T}}(P)) = \omega\, f(P) \quad , \tag{2.19}$$

so we have

$$|f(\tilde{\mathcal{T}}(P))| = |f(P)| \quad . \tag{2.20}$$

Thus $|f(P)|$ is invariant under $\tilde{\mathcal{T}}$. On the other hand (Sec. 1), a system has been shown to be ergodic if and only if any invariant measurable function is constant almost everywhere. Therefore we have the following property: if the dynamical system defined by $\tilde{\mathcal{T}}$ and μ on Φ is ergodic, then by (2.20), the absolute value of every eigenfunction of the induced operator $\mathcal{U}$ is constant (except possibly for a subset of measure zero). Notice that every eigenvalue ω must be simple. Indeed if g is any other eigenfunction of $\mathcal{U}$ with eigenvalue ω, since $f \neq 0$ by the discussion above and moreover by definition

$$\mathcal{U}\left(\frac{g}{f}\right) \equiv \frac{\mathcal{U}g}{\mathcal{U}f} = \frac{\omega g}{\omega f} = \frac{g}{f} \quad , \tag{2.21}$$

and g/f is an invariant function, then g cannot be but a constant multiple of f. On the other hand, if the eigenvalue of g is $\omega' \neq \omega$, then

$$\mathscr{U}\frac{g}{f} = \frac{\omega' g}{\omega f} = (\omega'\omega^{-1})\ \frac{g}{f} \tag{2.22}$$

i.e. g/f is an eigenfunction of $\mathscr{U}$ with eigenvalue $\omega'' = \omega'\omega^{-1} = e^{i(\tau'-\tau)}$. We have then also the statement: the system is ergodic if and only if 1 is a simple eigenvalue of the induced operator $\mathscr{U}$: i.e. $\exists$, $h \in \mathscr{H}$ such that $\mathscr{U}h = h$ (see (2.21)).

Finally let us rewrite the definition (1.62) in this form: the dynamical system defined by $\tilde{\mathscr{T}}$, μ and Φ is mixing if, for any pair of functions $f, g \in \mathscr{H}$,

$$\lim_{n\to\infty} \langle \mathscr{U}^n f|g\rangle = \langle f|1\rangle\langle 1|g\rangle \quad . \tag{2.23}$$

The equivalence is straightforward (identifying n with t and $\mu \equiv \mu_E$, $\Phi \equiv \mathscr{E}$, $\mathscr{U}^n f(P) = f(\tilde{\mathscr{T}}^n(P)) \equiv f(G(t|0)P))$, and if f, g are assumed to be the characteristic functions of the subsets $\mathscr{A}$ and $\mathscr{B}$ respectively of Φ leads to

$$\lim_{n\to\infty} \mu[\tilde{\mathscr{T}}^n \mathscr{A} \cap \mathscr{B}] = \mu(\mathscr{A})\ \mu(\mathscr{B}) \tag{2.24}$$

as expected by the general definition. Now, assuming that in (2.23) $f = g$ is equal to an eigenfunction of $\mathscr{U}$, with eigenvalue ω, we have

$$\lim_{n\to\infty} \langle \mathscr{U}^n f|f\rangle = \lim_{n\to\infty} \omega^n = |\langle f|1\rangle|^2 = \text{constant} \quad . \tag{2.25}$$

Hence $\omega = 1$, and one gets: the system is mixing if the only eigenvalue of $\mathscr{U}$ is 1. Incidentally, the property that mixing implies ergodicity is here recovered quite naturally.

The dynamical system $(\tilde{\mathscr{T}}, \mu, \Phi)$ is said to have a Lebesgue spectrum indexed in the set $\mathscr{I}$ if there exists an orthonormal basis $\mathbb{D} \equiv \{\psi_{i,j}|\ i \in \mathscr{I},\ j \in \mathbb{Z}\}$ of $\mathscr{H}$, including 1, and such that

$$\mathscr{U}\psi_{ij} = \psi_{i,j+1} \qquad \forall i,j \in \mathbb{Z} \ . \tag{2.26}$$

The cardinality of the set $\mathscr{I}$ is called the multiplicity of the Lebesgue spectrum. Notice the analogy of (2.26) with (7.26) of Chap. 1, which indeed could be used to show that the Bernoulli shift has a (countable) Lebesgue spectrum. Returning to the automorphism $\tilde{\mathscr{T}}$ of (2.2), and introducing first as orthonormal basis of $\mathscr{H}$ the set

$$\mathbb{B} \equiv \{\phi_{m,\ell}(q,p) = \exp(2\pi i(mq + \ell p)) | m,\ell \in \mathbb{Z}\} \qquad , \tag{2.27}$$

one immediately notices that $\mathscr{U}$ (see definitions (2.3) and (2.17)) induces on $\mathbb{B}$ the automorphism

$$\sigma : \begin{pmatrix} m \\ \ell \end{pmatrix} \longrightarrow \mathscr{T}\begin{pmatrix} m \\ \ell \end{pmatrix} = \begin{pmatrix} m + \ell \\ m + 2\ell \end{pmatrix} \qquad . \tag{2.28}$$

It was already pointed out that $\mathscr{T}$ has an eigenvalue $\tau_1 > 1$ and the other, $\tau_2 < 1$. Hence the orbit of $(m,\ell) \in \mathbb{Z}^2$, which is a bounded subset of $\mathbb{R}^2$ invariant under $\mathscr{T}$ is finite only if $m = \ell = 0$. Otherwise $\mathscr{T}$ is <u>dilating</u> in the proper direction corresponding to τ_1 and <u>contracting</u> in the proper direction corresponding to τ_2.

Thus $\mathbb{Z}^2 = \{0,0\}$ splits into a set $\mathscr{T}$ of orbits of σ, and correspondingly $\mathbb{B} - \{\phi_{0,0}\} : (\phi_{00} = 1)$ splits into a set of orbits $\{O_i\}$ of $\mathscr{U}$, labelled by the elements $i \in \mathscr{I}$. Each orbit on the other hand is obviously in one-to-one correspondence with $\mathbb{Z}$, in that it is generated by iteration (σ^n or $\mathscr{U}^n$ respectively). Denoting by $\psi_{i,0}$ some element of O_i, one may write

$$O_i = \{\psi_{i,n} = \mathscr{U}^n \, \psi_{i,0} \, | n \in \mathbb{Z}\} \tag{2.29}$$

from which it is obvious that $\psi_{i,n+1} = \mathscr{U}\psi_{i,n}$, as required by (2.26). Now, if one selects in (2.23) $f = \psi_{i,j}$, $g = \psi_{\ell,m}$, both orthogonal to 1, (2.23) itself reads

$$\lim_{n\to\infty} < \mathscr{U}^n \, \psi_{ij} | \psi_{\ell m} > = \lim_{n\to\infty} < \psi_{i,j+n} | \psi_{\ell m} > =$$

$$= < \psi_{ij} | 1 > < 1 | \psi_{\ell m} > = 0 \quad . \tag{2.30}$$

Thus $< \psi_{i,j+n} | \psi_{\ell m} >$ is certainly null for n large enough, and there follows that a dynamical system with Lebesgue spectrum is certainly mixing. Equation (2.29) then implies that (2.1), (2.2) define a system which is mixing.

This proof is particularly interesting in that for that system even ergodicity is difficult to prove on the basis of the general definition (1.54). Indeed time average and average over Φ are not everywhere coincidental. The line γ in the covering plane $\mathbb{R}^2$ of Φ

$$p = p_0 + (\tau_2 - 1)(q - q_0) \tag{2.31}$$

orthogonal to the solution of (2.4) corresponding to the initial data (q_0, p_0), projects onto the helix $\tilde{\gamma} \in \Phi$ under the natural projection $\mathbb{R}^2 \to \Phi$ obtained by identification of the opposite sides of the square $Q \equiv I \times I \subset \mathbb{R}^2$.

Once more by Jacobi's theorem, $\tilde{\gamma}$ is dense on Φ. On the other hand, let $P \equiv (q,p)$ be a point of $\tilde{\gamma}$ and we have

$$\tilde{\mathscr{T}}^n(P) = (q - q_0) \begin{pmatrix} a_n + (\tau_2 - 1) b_n \\ \\ c_n + (\tau_2 - 1) d_n \end{pmatrix} \pmod 1 \xrightarrow[n\to\infty]{} \begin{pmatrix} 0 \\ \\ 0 \end{pmatrix} \tag{2.32}$$

where we used (2.5), (2.10) - (2.12), together with the property (see (2.14)) that $(\tau_1 - 1)(\tau_2 - 1) = -1$. Considering now the analytic function

$$\psi(q,p) = \exp(2\pi i q) \tag{2.33}$$

we have — by (2.32) —

$$\bar{\psi} = \lim_{N \to \infty} \frac{1}{N} \sum_{n=0}^{N-1} \psi(\tilde{\mathscr{T}}^n(P)) = 1 \tag{2.34}$$

whereas

$$<\psi> = \int_{\Phi} \psi(q,p)dq\ dq = \int_0^1 e^{i2\pi q}dq = 0 \quad . \tag{2.35}$$

In other words, whatever the point P is on the dense subset $\tilde{\gamma}$, $<\psi> \neq \bar{\psi}$.

Consider now the tangent bundle $T\Phi$ of Φ, denote by $\| X \|$ the length of a tangent vector X to Φ referred to the Riemannian metric $ds^2 = dq^2 + dp^2$, and let $\mathscr{T}^{\#} : T\Phi_P \to T\Phi_{\tilde{\mathscr{T}}(P)}$ be the differential of $\tilde{\mathscr{T}}$. It is obvious that in the local chart (q,p) at P, $\mathscr{T}^{\#}$ is represented by $\mathscr{T}$ ((2.3)). Hence $\mathscr{T}^{\#}$ is dilating in the direction X corresponding to eigenvalue $\tau_1 > 1$ and contracting in the direction corresponding to eigenvalue $0 < \tau_2 < 1$. More precisely

i) $T\Phi_P$ is factorized into the direct sum of two subspaces

$$T\Phi_P = X_P \oplus Y_P \tag{2.36}$$

and

ii) there exist real constants a, b, λ such that for every positive integer n

$$\left.\begin{aligned} \| (\mathscr{T}^n)^{\#}\, \xi \| &\geq a \cdot e^{n\lambda}\ \| \xi \| \\ \| (\mathscr{T}^{-n})^{\#}\, \xi \| &\leq b \cdot e^{-n\lambda}\ \| \xi \| \end{aligned}\right\} \text{ if } \quad \xi \in X_P \tag{2.37}$$

and

$$\left.\begin{aligned} \| (\mathscr{T}^n)^{\#}\, \eta \| &\leq b \cdot e^{-n\lambda}\ \| \eta \| \\ \| (\mathscr{T}^n)^{\#}\, \eta \| &\geq a \cdot e^{n\lambda}\ \| \eta \| \end{aligned}\right\} \text{ if } \quad \eta \in Y_P \quad . \tag{2.38}$$

These two properties which characterize the instability of orbits mentioned at the beginning of this section hold in general (in the case under consideration $a = b = 1$, $e^{\lambda} = \tau_1$, $e^{-\lambda} = \tau_2$), for any metric g (in that case a and b depend on g) and define the class of ergodic mixing, highly unstable dynamical systems referred to as "Anosov".

The example of Smale and Arnold can finally be constructed. Let $T^2 \equiv \Phi = \{(q,p) \bmod 1\}$ be the two-dimensional torus, and $I = \{z | 0 \le z \le 1\}$ the unit interval. Construct first the cylinder $T^2 \times I$, and then identify $T^2 \times \{0\}$ with $T^2 \times \{1\}$ according to

$$((q,p),1) \equiv (\tilde{\mathscr{T}}(q,p),0) \tag{2.39}$$

where $\tilde{\mathscr{T}}$ is defined in (2.2). One obtains thus a compact manifold $\mathscr{M}$. The mapping π of $\mathscr{M}$ onto the unit circle $S^1 = I$ with extreme points identified $(S^1 = z\{(\bmod 1)\})$,

$$\pi : \mathscr{M} \to S^1 \quad , \quad \pi(q,p,z) = z \tag{2.40}$$

leads to identify $\mathscr{M}$ as a fibre bundle over S^1 with fibre T^2. π has rank 1. [Recall that a <u>fibre bundle</u> M over an n-dimensional smooth manifold B, called the <u>base</u>, consists of the triple objects (M, B, π), where M is a compact, connected, smooth (n+s)-dimensional manifold and π, called <u>projection</u>, is a (differentiable) mapping $\pi : M \to B$ of rank n. The sections $\pi^{-1}(b)$, $b \in B$, which are s-dimensional manifolds, diffeomorphic one to the other, are called the <u>fibres</u>].

Define the Riemannian metric g on $\mathscr{M}$

$$ds^2 = \tau_1^{2z}[\tau_1 dq + (1-\tau_1)dp]^2 + \tau_2^{2z}[\tau_2 dq + (1-\tau_2)dp]^2 + dz^2 \ . \tag{2.41}$$

g leads to an invariant measure over $\mathscr{M}$, in that it is invariant under the substitutions

$$q \to q + p \quad , \quad p \to q + 2p \quad , \quad z \to z + 1 \tag{2.42}$$

used to perform the identifications defining $\mathcal{M}$. The flow generated by the equations

$$\dot{q} = 0 \quad , \quad \dot{p} = 0 \quad , \quad \dot{z} = 1 \tag{2.43}$$

is Anosov.

Indeed, if $P \equiv (q,p,z) \in \mathcal{M}$, define the subspaces of $T\mathcal{M}_P$: X_P, tangent to the fibre $T^2 \times S^1$ and parallel to the proper direction of

$$\mathbb{T} : \; T^2 \times S^1 \to T^2 \times S^1$$

$$\mathbb{T} : \; \begin{pmatrix} q \\ p \\ z \end{pmatrix} \to \begin{pmatrix} 1 & 1 & 0 \\ 1 & 2 & 0 \\ 0 & 0 & 1 \end{pmatrix} \begin{pmatrix} q \\ p \\ z \end{pmatrix} \pmod 1 \tag{2.44}$$

corresponding to eigenvalue τ_1; Y_P, defined in the same way but in correspondence to the eigenvalue τ_2; Z_P, collinear to the vector $\begin{pmatrix} 0 \\ 0 \\ 1 \end{pmatrix}$, defining the velocity of the flow (2.43). Condition i) is thus obviously satisfied:

$$T\mathcal{M}_P = X_P \oplus Y_P \oplus Z_P \quad . \tag{2.45}$$

On the other hand, in the local chart (q,p,z) the vectors $\xi \in X_P$ are of the form $\xi = (t, (\tau_1 - 1)t, 0)$, (compare (2.4) and (2.43)) whereas the matrix of $\mathbb{T}^{\#}$ reduces to $\mathbb{T}$ (diagonal, eigenvalue τ_1). Using the metric (2.41) to compute vector lengths, one finds then

$$\| \mathbb{T}^{\#} \xi \| = \tau_1^t \, \| \xi \| \quad . \tag{2.46}$$

In the same way one obtains, for $\eta \in Y_P$

$$\| \mathbb{T}^{\#} \eta \| = \tau_2^t \, \| \eta \| = \tau_1^{-t} \, \| \eta \| \quad . \tag{2.47}$$

Conditions ii) are also satisfied, once more with $a = b = 1$, $\lambda = \ln \tau_1$.

On $X_p \oplus Z_p$ and $Y_p \oplus Z_p$, the system is smooth and completely integrable and defines a complete set of union of orbits which are asymptotic to each other as $t \to -\infty$ or $+\infty$ respectively.

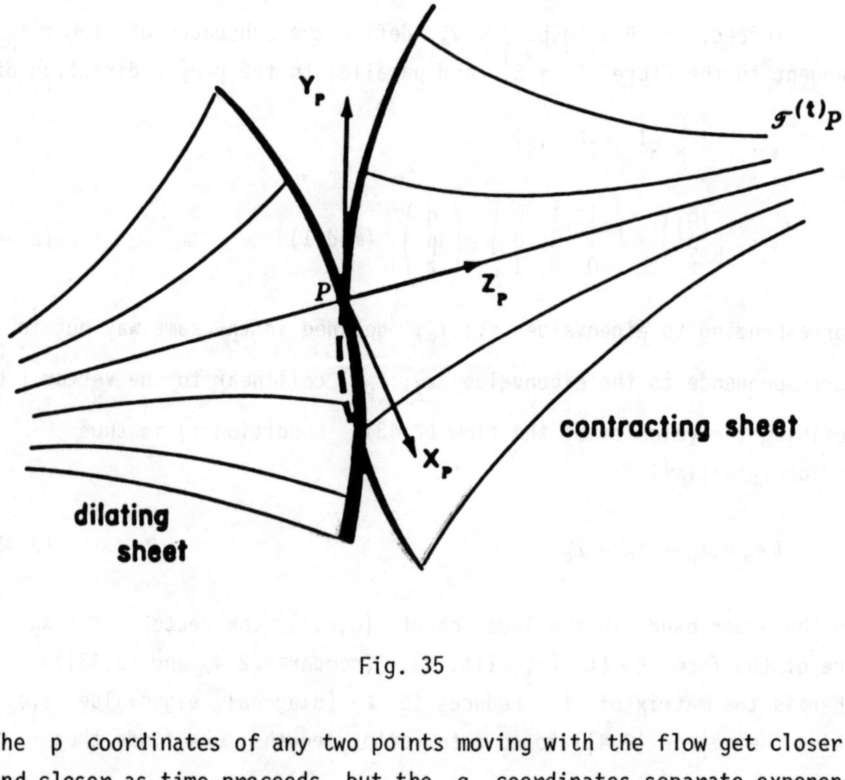

Fig. 35

The p coordinates of any two points moving with the flow get closer and closer as time proceeds, but the q coordinates separate exponentially fast — unless they were identical to start with — and hence the two points eventually move apart exponentially fast in phase space.

The physical reason why trajectories exhibiting the type of instability discussed so far lead to mixing is therefore that even a very small change in the initial conditions brings about a very large and practically unpredictable change in the subsequent motion. Even if the initial phase space density is concentrated in a very small region of the phase space, the ultimate behaviour will become undistinguishable from that of an equilibrium ensemble with density $\tilde{\rho}$.

It is worth noticing that Anosov dynamical systems are structurally stable in the sense described in Sec. 8 of Chap. 1, (the proof is based on the intuitive fact that given an automorphism $\tilde{\mathscr{T}}'$ in the neighbourhood of $\tilde{\mathscr{T}}$, then one can find a homeomorphism $\mathscr{X}$ of Φ such that

$$\tilde{\mathscr{T}}' = \mathscr{X}\mathscr{T}\mathscr{X}^{-1} \tag{2.48}$$

and by denoting $d(P,P')$ the Riemannian distance in Φ between the points $P, P' \in \Phi$, then — due to the structure shown in Fig. 35, characteristic of Anosov flows — for any $P \in \Phi$ and any $\tilde{\mathscr{T}}'$ in the neighbourhood of $\tilde{\mathscr{T}}$, $d(\mathscr{X}(P),P)$ is bounded). Then, in the presence of a small external perturbation, every trajectory of the perturbed system is close to a trajectory of the unperturbed system: thus the effect of any small external perturbation can be replaced by a small change of the initial conditions and hence be allowed for by means of the Gibbs ensemble method.

3. Connection with Thermodynamics

Statistical mechanics in its original set-up, was intended to be the basic theory underlying that phenomenological theory which is thermodynamics (as electromagnetics theory is for optics). As such it needs to deal with systems which are macroscopic in size, and therefore contain a very large number of particles.

There are two ways in statistical mechanics of taking into account that a system is very large on the molecular scale. One is the use of the thermodynamic limit, in which all calculations are made in terms of fixed values of the number of particles N, energy E and volume V, which are eventually allowed to become very large, in such a way that their ratios approach or maintain a finite given value. The other is to deal from the start with an infinitely extended system, in which the specific densities such as the number of particles or energy per unit volume are fixed.

The advantage of the thermodynamic limit method is that it demonstrates quite clearly that the laws of thermodynamics hold and justify the usual recipes of elementary statistical mechanics for calculating thermodynamic functions. The infinite system approach has the advantage that it eliminates the difficulties connected with the way one takes the limit $V \to \infty$. This is usually done by assuming periodic boundary conditions, which induces, when dealing with quantum systems, spurious discrete spectra. Moreover it allows writing equations describing macroscopic processes that are asymmetrical under time reversal (and can thus be assumed to describe irreversible processes) making a clear-cut distinction between the macroscopic and microscopic dynamics (which is always symmetric under time reversal) as microscopic dynamics do not determine completely the ensemble one should use.

For a classical infinite system, the phase space is infinite-dimensional, so the Gibbs ensemble cannot be described in terms of a phase space density. One resorts instead to the concept of <u>state</u>. The latter is a rule which associates to each local dynamical variable F a real number $\langle F \rangle$, called its expectation. Thus a state is a function (usually linear) from the space of local dynamical variables to the space of complex numbers. Notice that the definition of $\langle F \rangle$ used for finite systems is consistent with the above scheme.

However, an ensemble representing an infinite system should be described by local correlations instead of a density operator. We will not further elaborate these notions.

Instead, returning to the connection of statistical mechanics with thermodynamics, we will recall here only how it is essentially based on the concept of entropy, a concept which may well be extended to arbitrary dynamical systems — due to its information theoretic meaning (see Sec. 8 of Chapter 3).

The second law of thermodynamics ensures the existence of entropy, an additive thermodynamic function defined at equilibrium, and which can increase but never decrease in a process where no energy (heat) enters or leaves the system. Since it has to be a thermodynamic function (or a function of the thermodynamic state), the entropy S can be assumed, tentatively, as a function of $\rho(P)$, of the form

$$S_f\{\rho\} \int_{\Phi} f(\rho(P))d\Omega \tag{3.1}$$

where f is a function to be chosen. It is a consequence of Liouville's theorem that definition (3.1) implies

$$S_f\{\rho_{t,0}\} = S_f\{\rho_0\} \quad . \tag{3.2}$$

In order to give entropy the desired monotonicity property, one wants to choose f so that $S_f\{\rho_0\} \leq S_f\{\tilde{\rho}\}$. This can obviously be done if f is a concave function $(f'' < 0)$. Indeed, if f is concave

$$f(\rho_{t,0}(P)) \leq f(\tilde{\rho}(P)) + (\rho_{t,0}(P) - \tilde{\rho}(P))\, f'(\tilde{\rho}(P)) \tag{3.3}$$

whence by (1.42),

$$\lim_{T\to\infty} \frac{1}{T}\int_0^T f(\rho_{t,0}(P))dt \leq f(\tilde{\rho}(P)) \tag{3.4}$$

and furthermore

$$\lim_{T\to\infty} \frac{1}{T}\int_0^T S_f\{\rho_{t,0}\}dt \leq S_f\{\tilde{\rho}\} \quad . \tag{3.5}$$

But, due to (3.2), (3.5) reads as required

$$S_f\{\rho_0\} \leq S_f\{\tilde{\rho}\} \tag{3.6}$$

and the candidate entropy (3.1) is indeed such that the probability density $\tilde{\rho}$ which describes the final equilibrium has higher entropy than the density describing the initial state. Notice that (3.6) implies the non-interchangeability of the limit $t \to \infty$ with the functional in S_f. Concavity is not sufficient to determine f completely, however the thermodynamic entropy should be additive as well, namely the entropy of a system consisting of two independent subsystems is the sum of the entropies of the subsystems. This almost completely determines f.

Let $\rho_a(P)$, $\rho_b(P)$, $P \in \Phi$, be the probability densities for subsystems $\mathscr{A}$ and $\mathscr{B}$ respectively. Then

$$\rho(P,P')d\Omega d\Omega' = \rho_a(P)\ \rho_b(P')d\Omega d\Omega' \quad , \quad P,P' \in \Phi \quad , \tag{3.7}$$

is the probability that whereas the subsystem $\mathscr{A}$ is represented by a point in the phase space volume element $d\Omega$ at P, subsystem $\mathscr{B}$ is represented by a point in $d\Omega'$ at P'. If ρ_a, ρ_b are normalized, $\rho(P,P')$ is also normalized:

$$\int_\Phi \rho_c(P)d\Omega = 1\ , \quad c = a,b \rightarrow \int_{\Phi\times\Phi} \rho(P,P')d\Omega d\Omega' = 1 \quad . \tag{3.8}$$

The entropy for the composed system reads

$$\int_{\Phi\times\Phi} f(\rho(P,P'))d\Omega d\Omega' = \int_{\Phi\times\Phi} f(\rho_a(P)\ \rho_b(P'))d\Omega d\Omega' \quad . \tag{3.9}$$

We require that this should be equal to

$$\int_\Phi f(\rho_a(P))d\Omega + \int_\Phi f(\rho_b(P))d\Omega =$$

$$\int_{\Phi\times\Phi} \{\rho_b(P')f(\rho_a(P)) + \rho_a(P)f(\rho_b(P'))\}d\Omega d\Omega' \tag{3.10}$$

where the normalization of ρ_a,ρ_b was used. In other words the function f should satisfy the functional equation

$$f(xy) = xf(y) + yf(x) \tag{3.11}$$

setting first

$$f(x) = x \cdot g(x) \tag{3.12}$$

then, for $x, y \neq 0$,

$$g(xy) = g(x) + g(y) \quad . \tag{3.13}$$

The unique solution to (3.13) is

$$g(x) = -\,k\,\ell n\,x \tag{3.14}$$

which determines the entropy up to a positive constant k (which physically just determines the units and is the same for all systems). Notice that

$$f(x) = -\,kx\,\ell n\,x \tag{3.15}$$

has the required concavity property for $x > 0$,

$$f'' = -\frac{k}{x} < 0 \quad . \tag{3.16}$$

Inserting in (3.1), we have finally

$$S = -\,k \int_{\Phi} \rho(P)\,\ell n[\rho(P)]\,d\Omega \quad . \tag{3.17}$$

This identifies in a straightforward manner S with the function H defined in (1.46), (2.3) of Chapter 5 $(X \equiv \Phi)$ and allows the concept to extend to abstract dynamical systems. Let $\alpha = \{\mathscr{A}_i\}_{i \in \mathscr{I}}$ ($\mathscr{I}$ a finite or countable set) be a collection of non-empty, non-intersecting, μ-measurable sets which cover Φ:

$$\mu(\mathscr{A}_i \cap \mathscr{A}_j) = 0 \quad \text{if} \quad i \neq j \quad , \quad \mu(\Phi - \bigcup_{i \in \mathscr{I}} \mathscr{A}_i) = 0 \quad . \tag{3.18}$$

α is called a partition of Φ. A partition β is called a refinement of the partition α (denoted $\alpha \dot{\leq} \beta$), if every element $\mathscr{B}_k$ of β is a subset of some element $\mathscr{A}_k$ of α

$$\mu(\mathscr{B}_k - \mathscr{B}_k \cap \mathscr{A}_k) = 0 \quad \forall k \in \mathscr{I} \quad . \tag{3.19}$$

Let $\{\alpha_j\}_{j \in \mathscr{J}}$ be a collection of partitions. Define their sum

$$\tilde{\alpha} = \bigvee_{j \in \mathscr{J}} \alpha_j \tag{3.20}$$

as the "smallest" (with respect to $\dot{\leq}$) partition which contains every α_j, namely

$$\tilde{\alpha} = \{ \bigcap_{j \in \mathscr{J}} \{\mathscr{A}_i\}_j | \{\mathscr{A}_j\}_j \in \alpha_j \quad , \quad \forall j\} \quad . \tag{3.21}$$

The function

$$H(\alpha) = \sum_{i \in \mathscr{I}} \zeta\Big(\mu(\mathscr{A}_i)\Big) \tag{3.22}$$

where

$$\zeta(t) = \begin{cases} - t \operatorname{Log}_2 t & \text{if} \quad 0 < t \leq 1 \\ 0 & \text{if} \quad t = 0 \end{cases} \tag{3.23}$$

is referred to as the entropy of the partition α.

Let finally a dynamical system be defined on $\{\Phi,\mu\}$ by the automorphism $\tilde{\mathscr{T}}$ of Φ. We define as entropy of α relative to $\tilde{\mathscr{T}}$ the quantity

$$H_{\tilde{\mathscr{T}}}(\alpha) = \lim_{n \to \infty} \frac{1}{n} H\Big(\bigvee_{k=0}^{n-1} \tilde{\mathscr{T}}^k(\alpha)\Big) \tag{3.24}$$

and from it the notion — due to Kolmogorov — of entropy of the dynamical system

$$H = \sup_{\alpha} H_{\tilde{\mathcal{T}}}(\alpha) \tag{3.25}$$

where the supremum is taken over all finite partitions of Φ. H, which is clearly ≥ 0, is an invariant of the dynamical system. Indeed, consider an isomorphism $\mathcal{X}: \Phi \to \Phi'$, and construct over Φ' the dynamical system characterized by the map $\tilde{\mathcal{T}}' = \mathcal{X}\tilde{\mathcal{T}}\mathcal{X}^{-1}$ and measure $\mu' = \mu$. If α is a partition of Φ, $\alpha' = \mathcal{X}\circ\alpha$ will be a partition of Φ' but then, by (3.24)

$$\begin{aligned} H_{\tilde{\mathcal{T}}'}(\alpha') &= H_{\mathcal{X}\tilde{\mathcal{T}}\mathcal{X}^{-1}}(\mathcal{X}\circ\alpha) = \\ &= \lim_{n\to\infty} \frac{1}{n} H\,(\mathcal{X}\circ\alpha \vee \ldots \vee \mathcal{X}\tilde{\mathcal{T}}^{n-1}\mathcal{X}^{-1}\circ\mathcal{X}\circ\alpha) = \\ &= \lim_{n\to\infty} \frac{1}{n} H\,(\mathcal{X}\circ(\alpha \vee \ldots \vee \tilde{\mathcal{T}}^{n-1}(\alpha))) = H_{\tilde{\mathcal{T}}}(\alpha) \end{aligned} \tag{3.26}$$

where we used the definitions (3.24) as well as (3.22), which implies — due to the invariance of the measure μ — that $H(\mathcal{X}\circ\alpha) = H(\alpha)$. On the other hand, as α runs over all the partitions of Φ, $\mathcal{X}\circ\alpha$ runs over all the partitions of Φ' and

$$\sup_{\alpha'} H_{\tilde{\mathcal{T}}'}(\alpha') = \sup_{\alpha} H_{\tilde{\mathcal{T}}}(\alpha) = H \quad . \tag{3.27}$$

Finally we have the Kolmogorov theorem. If the partition $\tilde{\alpha}$ of Φ is such that

$$\bigvee_{k=-\infty}^{+\infty} \tilde{\mathcal{T}}^{k}(\tilde{\alpha}) = \varepsilon \tag{3.28}$$

where ε is the partition of Φ into points, i.e., the most refined partition possible (and the equality is understood mod-zero), $\tilde{\alpha}$ is called a generating partition for $\tilde{\mathcal{T}}$. Kolmogorov's theorem states that

$$H = H_{\tilde{\mathcal{T}}}(\tilde{\alpha}) \tag{3.29}$$

and thus enables to compute the entropy explicitly.

The proof of (3.29) is based on the observation that, if one defines for a given partition α,

$$\alpha_n = \bigvee_{k=-n}^{n} \tilde{\mathcal{T}}^k(\alpha) \tag{3.30}$$

then

$$\bigvee_{n=0}^{\infty} \tilde{\alpha}_n = \varepsilon \qquad . \tag{3.31}$$

Moreover

$$\tilde{\alpha}_1 \dot{\leq} \tilde{\alpha}_2 \dot{\leq} \ldots \dot{\leq} \tilde{\alpha}_n \leq \ldots \tag{3.32}$$

and, due to the invariance of μ and (3.24),

$$\frac{1}{n} H(\alpha_n) \xrightarrow[n\to\infty]{} 2H_{\tilde{\mathcal{T}}}(\alpha) \qquad . \tag{3.33}$$

The theorem then follows if one can prove that $\beta \dot{\leq} \tilde{\alpha}_n$ for some n implies $H_{\tilde{\mathcal{T}}}(\beta) \leq H_{\tilde{\mathcal{T}}}(\tilde{\alpha})$. Actually, if $\beta \dot{\leq} \tilde{\alpha}_n$, then

$$\beta_m \dot{\leq} (\tilde{\alpha}_n)_m = \tilde{\alpha}_{n+m} \tag{3.34}$$

for all m, hence

$$\frac{1}{m} H(\beta_m) \leq \frac{1}{m+n} H(\tilde{\alpha}_{n+m}) \frac{m+n}{m} \qquad . \tag{3.35}$$

Thus, letting $m \to \infty$ and using (3.33), we have just

$$H_{\tilde{\mathscr{T}}}(\beta) \le H_{\tilde{\mathscr{T}}}(\tilde{\alpha}) \quad . \tag{3.36}$$

But by (3.25), obviously (3.35) implies (3.29).

Now H and S can be connected. To the (piece-wise continuous) map $\tilde{\mathscr{T}}$, one can associate the Ruelle-Perron-Frobenius operator $\mathscr{L}_f$, $f \in \mathscr{H}$, as defined by

$$(\mathscr{L}_f \phi)(P) = \sum_{P' \in \tilde{\mathscr{T}}^{-1}(P)} \phi(P')e^{-f(P')} \quad , \quad P,P' \in \Phi \tag{3.37}$$

where $\phi \in \mathscr{H}$. The Fredhold determinant (or Ruelle zeta function) of $\mathscr{L}_f$ reads

$$D(\lambda|\tilde{\mathscr{T}}) = \exp\left[- \sum_{n=1}^{\infty} \frac{\lambda^n}{n} Q_n(\mathscr{L}_f)\right] \tag{3.38}$$

where $\lambda \in \mathbb{R}$ and

$$Q_n(\mathscr{L}_f) = \sum_{Q \in \mathrm{Fix}(\tilde{\mathscr{T}}^n)} \exp\left\{ - \sum_{k=0}^{n-1} f(\tilde{\mathscr{T}}^k(Q)) \right\} \tag{3.39}$$

where $\mathrm{Fix}(\tilde{\mathscr{T}}^n)$ is the set of all fixed points of $\tilde{\mathscr{T}}^n$.

The apparently complicated definition is made clear recalling the identity valid for any operator L on $\mathscr{H}$,

$$\ln \det L = \mathrm{Tr} \ln L \quad . \tag{3.40}$$

From (3.40), upon setting $L = \mathbb{1} - \lambda\mathscr{L}$, one derives

$$D(\lambda) = \det(\mathbb{1} - \lambda\mathscr{L}) = \exp \mathrm{Tr} \ln (\mathbb{1} - \lambda\mathscr{L}) =$$

$$= \exp\left\{ - \sum_{n=1}^{\infty} \frac{\lambda^n}{n} \mathrm{Tr}(\mathscr{L}^n) \right\} \quad . \tag{3.41}$$

Thus Q_n is nothing but the trace of $\mathscr{L}^n$.

Let us introduce now in the theory a real parameter $\beta \geq 0$ — to be treated as a (formal) inverse temperature — and develop the usual thermodynamics introducing, instead of $Q_n(\mathscr{L}_f)$ as defined in (3.39),

$$\mathscr{Q}_n(\beta) \equiv \sum_{Q \in \mathrm{Fix}(\mathscr{T}^n)} \exp\left\{-\beta \sum_{k=0}^{n-1} f(\mathscr{T}^n(Q))\right\} \quad . \tag{3.42}$$

and assuming

$$F(\beta) = -\frac{1}{\beta} \lim_{n\to\infty} \frac{1}{n} \ln \mathscr{Q}_n(\beta) \tag{3.43}$$

as free energy per unit step. The corresponding entropy is

$$S(\beta) = \beta^2 \frac{\partial F}{\partial \beta} = \beta[U(\beta) - F(\beta)] \tag{3.44}$$

where $U(\beta)$ is the internal energy,

$$U(\beta) = \frac{\partial}{\partial \beta}(\beta F(\beta)) \quad . \tag{3.45}$$

We have, from (3.43), (3.44),

$$S(\beta) = \lim_{n\to\infty} \frac{1}{n}\left\{\ln \mathscr{Q}_n(\beta) - \beta \frac{\partial}{\partial \beta} \ln \mathscr{Q}_n(\beta)\right\} =$$

$$\equiv \lim_{n\to\infty} \frac{1}{n} S_n \quad . \tag{3.46}$$

Write now (3.42) in the form

$$\mathscr{Q}_n(\beta) = \sum_{i\in\mathscr{I}} \exp(-\beta E_i^{(n)}) \tag{3.47}$$

where $\mathscr{I}$ is a set in one-to-one correspondence with $\mathrm{Fix}(\mathscr{T}^n)$, and

$$E_i^{(n)} = \sum_{k=0}^{n-1} f(\tilde{\mathcal{T}}^k(P_i)) \quad , \quad i \in \mathcal{I} \leftrightarrow p_i \in \mathrm{Fix}(\mathcal{T}^n) \quad . \tag{3.48}$$

Upon defining the measurable sets $\mathcal{A}_i^{(n)}$, $i \in \mathcal{I}$, $n \in \mathbb{N}$, by

$$\mu(\mathcal{A}_i^{(n)}) = \frac{\exp(-\beta E_i^{(n)})}{\mathcal{Q}_n} \geq 0 \ , \qquad \sum_{i\in\mathcal{I}} \mu(\mathcal{A}_i^{(n)}) = 1 \tag{3.49}$$

— whereby using (3.48),

$$\ell n\, \mathcal{Q}_n = - \ell n\, \mu(\mathcal{A}_i^{(n)}) - \beta E_i^{(n)} \tag{3.50}$$

— it is straightforward to check that S_n, as defined in (3.46), writes

$$S_n = - \sum_{i\in\mathcal{I}} \mu(\mathcal{A}_i^{(n)})\, \ell n\, \mu(\mathcal{A}_i^{(n)}) =$$

$$= \ell n 2 \sum_{i\in\mathcal{I}} \zeta(\mu(\mathcal{A}_i^{(n)})) = \ell n 2\ H(\alpha) \tag{3.51}$$

where α is the partition corresponding to $\{\mathcal{A}_i^{(n)}\}_{i\in\mathcal{I}}$ and definition (3.22) was used. On the other hand, by (3.48), (3.30) α in (3.51) is given by

$$\alpha \equiv \alpha_{n-1} = \bigvee_{k=0}^{n-1} \tilde{\mathcal{T}}^k(\alpha) \tag{3.52}$$

and for the definition (3.49) if f is a non-negative function

$$\lim_{n\to\infty} \alpha_n = \tilde{\alpha} \quad . \tag{3.53}$$

Combining (3.46), (3.51) and (3.52) with (3.24), (3.53) and (3.29), we finally have

$$H = \mathrm{Log}_2 e \cdot S(\beta) \tag{3.54}$$

i.e., up to an irrelevant factor (due to the definition (3.23) of ζ), the Kolmogorov entropy equals the thermodynamic entropy.

Notice that if Φ $(\dim \Phi = d)$ is a lattice of n^d points, endowed with periodic boundary conditions (so that indeed Φ is a toroidal lattice), $\mathscr{T}$ is the mapping sending each point into its nearest neighbours and f is the interaction energy, — which for simplicity can be assumed to be confined to pairs of nearest neighbour points — then $S(\beta)$ is the canonical ensemble entropy, and $\mathscr{Q}_n(\beta)$ is the n-site partition function corresponding to the Boltzmann weight. Indeed, the latter is defined as the probability distribution which maximizes the entropy (3.17) subjected to the constraint that the system is in contact with a thermal reservoir, i.e. an energy density $\varepsilon(P)$, $P \in \Phi$ can be defined such that

$$\int_\Phi \rho(P)\ \varepsilon(P) d\Omega = U \tag{3.55}$$

is a constant (internal energy). Recalling that also the constraint (1.3) should be satisfied, one introduces two Lagrange multipliers α, β and the equation

$$\delta(S + \alpha + \beta U) = 0 \tag{3.56}$$

where δ denotes functional derivative with respect to ρ, readily gives

$$\delta\left(\int_\Phi \{\rho(P)\ell n\rho(P) + \alpha\rho(P) + \beta\varepsilon(P)\rho(P)\} d\Omega\right) =$$

$$= \int_\Phi \Big(1 + \ell n\rho(P) + \alpha + \beta\varepsilon(P)\Big)\ \delta\rho(P) d\Omega = 0 \tag{3.57}$$

i.e.,

$$\rho(P) = R \cdot e^{-\beta\varepsilon(P)} \quad . \tag{3.58}$$

The constant $R = e^{-(1+\alpha)}$ simply derives from the normalization condition (1.3) for $\rho(P)$ and thus defines the partition function

$$Z = R^{-1} = \int_{\Phi} e^{-\beta\varepsilon(P)} d\Omega \quad . \tag{3.59}$$

Upon defining an entropy density $s(P)$, by

$$S = \int_{\Phi} s(P)\rho(P)d\Omega \tag{3.60}$$

comparison of (3.60) and (3.17), taking (3.58) and (3.59) into account gives

$$-\frac{1}{k}s(P) = -\beta\varepsilon(P) - \ell nZ \quad . \tag{3.61}$$

The factor $-\ell nZ$ equals $\beta f(P)$, where $f(P)$ is now the free energy density. Recalling finally the thermodynamic definition of absolute temperature

$$T = \left(\frac{\partial s}{\partial \varepsilon}\right)^{-1} \tag{3.62}$$

one has

$$T = \frac{1}{k\beta} \tag{3.63}$$

which identifies the universal constant k in (3.17) with Boltzmann's constant k_B, so that

$$\rho(P) = \frac{e^{-\frac{\varepsilon(P)}{k_BT}}}{\int_{\Phi} e^{-\frac{\varepsilon(P)}{k_BT}} d\Omega} \tag{3.64}$$

to be compared with (3.49), (3.47). Assume now $\dim \Phi = 1$, and set in (3.42)

$$f(P) = \ln \| \mathrm{grad}_{P'} \tilde{\mathcal{T}}(P') \| \Big|_{P'=P} \quad . \tag{3.65}$$

It is straightforward to check by recursion and due to the chain rule of derivation that in this case if $Q \in \mathrm{Fix}(\tilde{\mathcal{T}}^n)$.

$$\sum_{k=0}^{n-1} f(\tilde{\mathcal{T}}^k(Q)) = \ln \| \mathrm{grad}_P \tilde{\mathcal{T}}^n(P) \| \Big|_{P=Q} \quad . \tag{3.66}$$

The quantity

$$\ell(P) = \lim_{n\to\infty} \frac{1}{n} \ln \| \mathrm{grad}_{P'} \tilde{\mathcal{T}}^n(P') \| \Big|_{P'=P} \tag{3.67}$$

is called the Lyapunov exponent of the mapping $\tilde{\mathcal{T}}$ at point $P \in \Phi$, and is usually assumed as the characteristic parameter measuring the exponential separation of adjacent points (typical of systems exhibiting chaos) under the action of a map. Equations (3.42), (3.43) and (3.66), (3.67) combined together give a different, intuitive explanation of the information theoretical content of the quantity $\mathcal{Q}_n$ and of the related thermodynamics.

As a final remark, notice that from (3.42) - (3.44) one gets, for $\beta = 0$,

$$S(0) = \lim_{n\to\infty} \frac{1}{n} \ln(\mathrm{ord}\ \mathrm{Fix}(\tilde{\mathcal{T}}^n)) \tag{3.68}$$

where ord denotes the order (i.e. the number of elements) of a set. Now the right-hand side of (3.68) is but the definition of the <u>topological entropy</u> of the mapping $\tilde{\mathcal{T}}$ on Φ. Recalling that, due to the positivity of the heat capacity

$$C = -\beta^2 \frac{\partial U}{\partial \beta} = -\beta \frac{\partial S}{\partial \beta} \quad . \tag{3.69}$$

S must be a decreasing function of β, so we have for all β,

$$S(0) \geq S(\beta) \tag{3.70}$$

i.e. the topological entropy is larger or at most equal to the Kolmogorov entropy. The latter theorem constitutes a statement of the Dinaburg's theorem.

Chapter 3
Elements of Probability Theory

1. Measures in a Vector Space

Let $\mathscr{V}$ be a finite dimensional vector space, $n = \dim \mathscr{V}$, $\mathscr{D}^{(0)}(\mathscr{V})$ the space of continuous functions over $\mathscr{V}$ with compact support and $\mathscr{D}_K^{(0)}(\mathscr{V})$, the space of continuous functions over $\mathscr{V}$ with support in $K \subset \mathscr{V}$, K compact. Any linear form over the space $\mathscr{D}^{(0)}(\mathscr{V})$,

$$\mu : f \to \int f \cdot \mu \quad , \quad f \in \mathscr{D}^{(0)}(\mathscr{V}) \tag{1.1}$$

whose restriction to $\mathscr{D}_K^{(0)}(\mathscr{V})$ is continuous for any K compact, is a <u>measure</u> over $\mathscr{V}$. Let B be a basis of $\mathscr{V}$, and denote for all $v \in \mathscr{V}$, by v_B the n-tuple of the components of v in B. In B, every function f is represented by a function f_B defined by

$$f_B(v_B) = f(v) \quad . \tag{1.2}$$

The condition (necessary and sufficient) for $f \in \mathscr{D}^{(0)}(\mathscr{V})$ is that of $f_B \in \mathscr{D}^{(0)}(\mathbb{R}^n)$. Then every measure μ on $\mathscr{V}$ is represented by a measure μ_B on $\mathbb{R}^n$ such that

$$\int f(v) \cdot \mu = \int f_B(v_B) \cdot \mu_B \quad . \tag{1.3}$$

μ_B is nothing but the image of μ by the mapping $v \to v_B$. In order that μ be bounded, it is sufficient that μ_B be bounded.

Conversely, if one associates a measure μ_B to every basis B, the necessary and sufficient condition for μ_B to be the representation of a measure μ on $\mathscr{V}$ is that

$$\int f_B(v_B) \cdot \mu_B \tag{1.4}$$

be independent of B, namely that for any change of basis defined by

$$v_{B'} = S^{-1} v_B \tag{1.5}$$

where S is the matrix performing the change, one has

$$\int f_B(Sv_{B'}) \cdot \mu_{B'} = \int f_B(v_B) \cdot \mu_B \quad . \tag{1.6}$$

In other words μ_B should be the image of $\mu_{B'}$ under the mapping $S : v_{B'} \to v_B$. The case of the Lebesgue measure is of special interest. Notice first that the Lebesgue measure in a vector space is defined only up to a multiplicative constant. Indeed let us set, in a particular basis B,

$$\mu_B = dv_B^1 \ldots dv_B^n \quad . \tag{1.7}$$

Considering the image of μ_B under the mapping S^{-1}, one gets

$$\mu_{B'} = |\det S| \; dv_{B'}^1 \ldots dv_{B'}^n \quad . \tag{1.8}$$

On the other hand, if $\mathscr{V}$ is euclidean (i.e., a finite dimensional Hilbert space), there exists a canonical Lebesgue measure dv over $\mathscr{V}$, defined by

$$\int f(v) dv = \int f_B(v_B) \sqrt{g(B)} \; dv_B^1 \ldots dv_B^n \quad , \tag{1.9}$$

where

$$g(B) = \det G(B) \tag{1.10}$$

with G(B) denoting the Gram matrix of the basis B, i.e. the matrix of elements

$$G_{ij} = < v_i \, , v_j > \quad . \tag{1.11}$$

[Notice that in (1.11) the bilinear form (scalar product) $b : \mathscr{V} \times \mathscr{V} \to \mathbb{R}$; $b(v,w) = <v,w> = \int v(\xi) w(\xi) \cdot \mu^{\xi}$, $v,w \in \mathscr{V}$ is simply related to the norm of $\mathscr{V}$,

$$\| v \| = \sqrt{<v,v>} \quad , \qquad v \in \mathscr{V} \quad . \tag{1.12}$$]

By the change of basis induced by S (Eq. (1.5)), one has indeed

$$g(B') = |\det S|^2 \, g(B) \tag{1.13}$$

which guarantees the invariance of the definition (1.9) with respect to the basis.

2. Probability Distributions

A finite dimensional vector space $\mathscr{V}$ is said to be equipped with a probability distribution and its elements are called possibilities, if a bounded positive measure $\tilde{\mu}$ is given in $\mathscr{V}$ normalized to unity, i.e.

i) $\int f \cdot \tilde{\mu} \geq 0 \quad$ for all $\quad f \in \mathscr{D}^{(0)}(\mathscr{V}) \quad , \quad f \geq 0$

ii) $\int 1 \cdot \tilde{\mu} = \int \tilde{\mu} = 1 \quad .$

Every measurable subset of A of $\mathscr{V}$ is called an event, and its measure — denoted prob(A) — is called the probability of the event A.

One says that the event A happens within the possibility a, $a \in \mathscr{V}$ if $a \in A$. The event A is impossible if A has measure zero, i.e. if $A = \emptyset$. Given two events A and B, the event $AB \doteq A \cap B$ is the event that happens if and only if both A and B happen. If $AB = \emptyset$, namely if AB is impossible, the two events A and B are said to be imcompatible. The event $A + B \doteq A \cup B$ is also an event, which happens if and only if one of the two events A or B happens. The two concepts above can be generalized in a straightforward way to any number of component events. The event $C(A)$, which happens if and only if A does not happen, is called the contrary of A.

Among the possible probability distributions, two play a special role in the applications.

a. Discrete distributions

They have the form

$$\sum_i p_i \, \delta_{a_i} \tag{2.1}$$

where $\{a_i\}$ is a numerable set of discrete points in $\mathscr{V}$ and where the real coefficients p_i satisfy the conditions

$$\sum_i p_i = 1 \quad , \quad p_i \geq 0 \quad \forall i \quad . \tag{2.2}$$

δ_a is the Dirac measure at point a, defined by

$$f(a) = \int f \cdot \delta_a \quad . \tag{2.3}$$

Due to (2.2) the coefficients p_i can be interpreted as the probabilities of the events a_i. In fact, due to (2.1) ÷ (2.3),

$$p_i = \text{prob}(a_i) \quad . \tag{2.4}$$

In (2.4) the possibilities a_i (points in $\mathscr{V}$) are identified with the events consisting of this possibility alone, namely with the events which may happen only if one particular possibility is realized.

If A is a proper event, i.e. a measurable part of $\mathscr{V}$,

$$\mathrm{prob}(A) = \sum_{a_i \in A} p_i \quad .$$

b. <u>Absolutely continuous distributions</u>

They are defined in terms of a probability density in relation with a Lebesgue measure,

$$\tilde{\mu}^v = \theta(v)dv \tag{2.5}$$

where $\theta \geq 0$, $\theta \in L^1$ (the space of integrable functions with respect to Lebesgue measure). One has

$$\mathrm{prob}(A) = \tilde{\mu}(A) = \int_A \theta(v)dv \quad . \tag{2.6}$$

As an example we can discuss briefly here one distribution of special physical interest; the centered Gaussian distribution in a vector space $\mathscr{V}$. Let us consider in $\mathscr{V}$ a definite positive quadratic form q, which in any basis B is represented as $q : v \to v_B^t \mathscr{A}(B) v_B$. Let us define — for every basis B — the measure

$$\mu_B = \frac{\sqrt{\det \mathscr{A}(B)}}{(2\pi)^{n/2}} \exp\left[-\frac{1}{2} v_B^t \mathscr{A}(B)\, v_B\right] dv_B^1 \ldots dv_B^n \quad . \tag{2.7}$$

Under the change of basis defined by $v_B = Sv_{B'}$, the image of μ_B is

$$\frac{\sqrt{\det \mathscr{A}(B)}}{(2\pi)^{n/2}} \exp\left[-\frac{1}{2} v_{B'}^t \mathscr{A}(B')\, v_{B'}\right] |\det S|\, dv_{B'}^1 \ldots dv_{B'}^n \quad . \tag{2.8}$$

Since

$$\mathscr{A}(B') = S^t \mathscr{A}(B) S \tag{2.9}$$

and hence

$$\det \mathscr{A}(B') = \det \mathscr{A}(B) \, |\det S|^2 \tag{2.10}$$

such image is indeed equal to μ_B. There exists therefore a measure $\tilde{\mu} \geq 0$ whose representation in every basis B is equal to μ_B. In order to show that $\int \tilde{\mu} = 1$, one needs then only to prove that $\int \mu_B = 1$ for a special choice of B. Let us choose B in such a way that $\mathscr{A}(B) = \mathbb{I}_n$. Then (we drop the index B at the right-hand side for simplicity)

$$\mu_B = \frac{1}{(2\pi)^{n/2}} \exp\left[-\frac{1}{2}\sum_i (v^i)^2\right] dv^1 \ldots dv^n \tag{2.11}$$

whence

$$\int \mu_B = \frac{1}{(2\pi)^{n/2}} \prod_i \int_{-\infty}^{+\infty} \exp\left[-\frac{1}{2}(v^i)^2\right] dv^i = 1 \quad . \tag{2.12}$$

$\tilde{\mu}$ is a probability distribution, referred to as the <u>centered Gaussian distribution on $\mathscr{V}$</u>.

We can now state the basic theorems and definitions of probability theory. The "theorem of total probabilities" goes as follows:- let $\mathscr{V}$ be a vector space equipped with a probability distribution and A an event sum of a denumerable family of pairwise incompatible events A_i. Then:

$$\text{prob}(A) = \sum_i \text{prob}(A_i) \quad . \tag{2.13}$$

The proof of (2.13) is simply obtained by recalling that $\text{prob}(A) = \mu(A) = \sum_i \mu(A_i)$ and $\mu(A_i) = \text{prob}(A_i)$. Straightforward corollaries of the theorem are:

i) $\text{prob}|C(A)| = 1 - \text{prob}(A) \quad ,$ (2.14)

ii) $\text{prob}(A+B) = \text{prob}(A) + \text{prob}(B) - \text{prob}(AB)$ (2.15)
for arbitrary events A and B .

The event A is said to be almost sure if prob(A) = 1. The event A is said to be almost impossible if prob(A) = 0. Two events A, B such that AB is almost impossible are said to be almost incompatible.

Let A be an event with non-vanishing probability. Consider the probability distribution

$$\tilde{\mu}_A = \frac{\varepsilon_A \tilde{\mu}}{\text{prob}(A)} \tag{2.16}$$

where ε_A is the characteristic function of the set A. Obviously then

$$\int \tilde{\mu}_A = 1 \quad . \tag{2.17}$$

For any event B, we define prob(B|A) the probability of B with respect to the probability distribution $\tilde{\mu}_A$, and refer to it as the conditional probability of B when A is realized. From (2.16) we have

$$\text{prob}(B|A) = \frac{\text{prob}(BA)}{\text{prob}(A)} \tag{2.18}$$

namely

$$\text{prob}(BA) = \text{prob}\,A \cdot \text{prob}(B|A) \quad . \tag{2.19}$$

The theorem of composed probabilities states:- let $\{A_i\}$ be a denumerable family of pairwise almost imcompatible events of a almost sure sum. For any event B, one has

$$\text{prob}(B) = \sum_i \text{prob}(A_i)\ \text{prob}(B|A_i) \quad . \tag{2.20}$$

The proof of the theorem stems from the fact that

$$\text{prob}(B) = \sum_i \text{prob}(BA_i) \tag{2.21}$$

together with (2.19).

3. Random Variables

Let $\mathscr{V}$ be a vector space equipped with a probability distribution. Any final variable (scalar or vectorial) of a measurable mapping over $\mathscr{V}$ is called a random variable. The quantity

$$E(x) = \int f(v) \cdot \tilde{\mu}^V \tag{3.1}$$

where $\tilde{\mu}^V$ is the probability distribution on $\mathscr{V}$, is called the average of the random variable $x = f(v)$. Notice that $E(xy) = \langle x, y \rangle$. The definition (3.1) is given with the following constraints:

a) if x is a scalar variable, $E(x)$ is defined for $f \in L^1$;

b) if x is a vector variable, assuming values in the vector space $\mathscr{W}$, (3.1) is to be read in the following sense: for all linear forms ℓ on $\mathscr{W}$, one has

$$E[\ell(x)] = \int \ell[f(v)] \cdot \tilde{\mu}^V \tag{3.2}$$

or, in a particular basis B,

$$E(x_B^i) = \int f_B^i(v) \cdot \tilde{\mu}^V \quad . \tag{3.3}$$

The formula (3.3) of course holds for any other basis B' as well. Then $E(x)$ is defined if $\ell \circ f \in L^1$ for all linear forms ℓ on $\mathscr{W}$ or, equivalently, if $f_B^i \in L^1$, for a special basis B and for all i's.

From definition (3.1) the following properties follow

$$E(\lambda_1 x_1 + \lambda_2 x_2) = \lambda_1 E(x_1) + \lambda_2 E(x_2) \tag{3.4}$$

where λ_1, λ_2 are constants and

$$E(a) = a \quad \text{if} \quad a = \text{constant} \quad . \tag{3.5}$$

Two random variables $x_1 = f_1(v)$, $x_2 = f_2(v)$ are said to be almost surely equal if one has $f_1(v) = f_2(v)$ almost everywhere with respect to the measure $\tilde{\mu}^V$, namely if the event $f_1(v) = f_2(v)$ is almost sure (the set in $\mathcal{V}$ where $f_1(v) \neq f_2(v)$ has measure zero with respect to $\tilde{\mu}^V$). In the case of a discrete distribution, $\tilde{\mu} = \sum_i p_i \delta_{a_i}$, one has

$$E(x) = \sum_i p_i f(a_i) = \sum_i p_i x^i \tag{3.6}$$

where $x^i = f(a_i)$. If the possibilities a_i are finite in number, say $i = 1,\dots,m$; one can identify the variable x with the m-tuple $(x^1,\dots,x^m)$, whose components are the values assumed by x at the possibilities a_i. The average $E(x)$ is then nothing but the average of such values computed with weights p_i. If m is infinite, of course $E(x)$ exists only if the series at the right-hand side of (3.6) is absolutely convergent.

In the case of an absolutely continuous distribution, as given by (2.5) one has

$$E(x) = \int f(v)\, \theta(v)\, dv \quad . \tag{3.7}$$

One can associate with any event A a random variable η_A, referred to as Bernoulli variable of the event, defined in the following way:

$$\eta_A(v) = \begin{cases} 1 & \text{if} \quad v \in A \\ 0 & \text{if} \quad v \notin A \end{cases} \quad . \tag{3.8}$$

η_A is therefore nothing but the characteristic function ε_A of the set A and one has

$$E(\eta_A) = \int_A \tilde{\mu} = \tilde{\mu}(A) = \text{prob}(A) \quad . \tag{3.9}$$

The variable v is said to be centered if $E(v) = 0$. Centered variables constitute a vector space, which is a subspace of the vector space L^1 of random variables with defined averages.

Let $\mathscr{V}$ be a vector space equipped with a probability density, $v \in \mathscr{V}$ and $x = f(v)$ a random variable over $\mathscr{V}$. Let $x \in \mathscr{W}$, $\mathscr{W}$ a vector space. The probability distribution $\tilde{\mu}^v$ over $\mathscr{V}$ admits an image $\tilde{\nu}^x$ over $\mathscr{W}$, induced by f, which is referred to as the probability distribution of x (notice that $\tilde{\nu}^x > 0$, $\int \tilde{\nu}^x = 1$).

Such an image is such that

$$\int g(x) \cdot \tilde{\nu}^x = \int g[f(v)] \cdot \tilde{\mu}^v \tag{3.10}$$

for all scalar functions g measurable with respect to $\tilde{\nu}^x$ (provided of course that both sides of (3.10) are defined). Upon setting $z = g(x) = g[f(v)]$, the variable z can be considered both as a random variable on $\mathscr{W}$ and as a random variable on $\mathscr{V}$, and then the two sides of (3.10) represent its average, defined on $\mathscr{W}$ and $\mathscr{V}$ respectively:

$$E_{\mathscr{W}}(z) = E_{\mathscr{V}}(z) \quad . \tag{3.11}$$

Thus — by the very definition of the image probability distribution $\tilde{\nu}$ — the average value of a random variable z can be indifferently computed on $\mathscr{W}$ by the distribution $\tilde{\nu}$ or on $\mathscr{V}$ by the distribution $\tilde{\mu}$. The common value of such an average will be simply denoted $E(z)$. These properties generalize in a straightforward way the case when z is a vector random variable.

4. Characterization of a Probability Distribution

Let us consider now the concepts most commonly utilized in the description or characterization of probability distributions. We shall confine our attention to the notions usually adopted in physical applications.

A) Let x be a scalar random variable, with probability distribution $\tilde{\mu}^x$. The function

$$F(\mathscr{x}) = \int_{]-\infty, \mathscr{x}[} \mu^x = \text{prob}\{x < \mathscr{x}\} \tag{4.1}$$

is called cumulative function. F is a monotonically increasing function of x, with $F(-\infty) = 0$, $F(+\infty) = 1$. Moreover $F(x)$ is continuous on the left, and

$$F(x+0) - F(x) = \mu(\{x\}) = \text{prob}(x = x) \quad . \tag{4.2}$$

The latter statements are easily proved with the observation that if $\{x_n\}$ is a sequence of real numbers, such that $x_{n+1} \geq x_n$, $x_n \leq x$ $\forall n$, and $\lim_{n \to \infty} x_n = x$, and one denotes by ε_n the characteristic function of the interval $]-\infty, x_n[$, by ε the characteristic function of the interval $]-\infty, x[$, one has

$$\varepsilon(x) = \lim_{n \to \infty} \varepsilon_n(x) \tag{4.3}$$

whence

$$F(x) = \int \varepsilon(x) \cdot \tilde{\mu}^X = \lim_{n \to \infty} \int \varepsilon_n(x) \cdot \tilde{\mu}^X =$$

$$= \lim_{n \to \infty} F(x_n) \quad . \tag{4.4}$$

If moreover $\{x_k\}$ is a sequence of real numbers such that $x_{k+1} \leq x_k$, $x_k \geq x$ $\forall k$, and $\lim_{k \to \infty} x_k = x$, one has

$$\lim_{k \to \infty} \varepsilon_k(x) = \begin{cases} 1 & \text{if} \quad x \in]-\infty, x[\\ \\ 0 & \text{if} \quad x \notin]-\infty, x[\end{cases} \tag{4.5}$$

whence

$$F(x+0) = \lim_{k \to \infty} F(x_k) = \lim_{k \to \infty} \int \varepsilon_k(x) \cdot \tilde{\mu}^X = \int_{]-\infty, x]} \tilde{\mu}^X =$$

$$= \tilde{\mu}(]-\infty, x[) + \tilde{\mu}(\{x\}) = F(x) + \tilde{\mu}(\{x\}) \tag{4.6}$$

which is indeed (4.2). Notice that if $\tilde{\mu}$ does not contain point Dirac measures, F is a continuous function; if $\tilde{\mu}$ is a finite sum of Dirac measures, then F is a staircase function.

Let us also recall that

$$\int_{[a,b[} \tilde{\mu}^x = F(b) - F(a) = \mathrm{prob}(a \leq x < b) \quad . \tag{4.7}$$

Denote by $\bar{x}$ the random variable $x - E(x)$. Obviously one has $E(\bar{x}) = 0$, in other words $\bar{x}$ is centered. Introduce then the following definitions:

i) moment of order k of the probability distribution:

$$m_k(x) \doteqdot E(x^k) \tag{4.8}$$

ii) centered moment of order k of the probability distribution:

$$\bar{m}_k(x) \doteqdot m_k(\bar{x}) \quad . \tag{4.9}$$

These moments can be defined for k up to some k_{max} (possibly ∞), depending on the probability distribution. One has

$$m_1(x) = E(x) \quad , \quad \bar{m}_1(x) = 0 \quad . \tag{4.10}$$

The centered moment of order 2 is called variance, and customarily denoted by $\sigma^2(x)$:

$$\bar{m}_2(x) \doteqdot \sigma^2(x) = E[(x - E(x))^2] = E(x^2) - [E(x)]^2 = \\ = m_2(x) - [m_1(x)]^2 \quad . \tag{4.11}$$

Notice that

$$\sigma^2(x + a) = \sigma^2(x) \quad \text{for all} \quad a \in \mathbb{R} \tag{4.12}$$

and

$$\sigma^2(cx) = c^2\sigma^2(x) \qquad , \qquad c \in \mathbb{R} \qquad . \tag{4.13}$$

The positive square root of the variance, $\sigma(x)$ is called <u>deviation</u>.

A random variable x is said to be <u>reduced</u> if $E(x) = 0$ and $\sigma(x) = 1$. Whatever the variable x, the variable $\bar{x}/\sigma(x)$, referred to as reduced deviation of x, is reduced. For an interesting example, consider the Bernoulli variable η_A of an event of probability p. Let $q = 1 - p$. Since, by definition, $\eta_A^2 = \eta_A$, one has

$$E(\eta_A^2) = E(\eta_A) = p \tag{4.14}$$

whence

$$\sigma^2(\eta_A) = p - p^2 = p(1-p) = pq \tag{4.15}$$

or

$$\sigma(\eta_A) = \sqrt{pq} \qquad . \tag{4.16}$$

One defines as <u>characteristic function</u> of the random variable x the Fourier transform Φ_x of the probability distribution $\tilde{\mu}^x$:

$$\Phi_x(s) = \int e^{isx} \cdot \tilde{\mu}^x = E(e^{isx}) \qquad . \tag{4.17}$$

Φ_x is a continuous function of s, with $\Phi_x(0) = 1$. If the moments $m_k(x)$ exist up to the order k_{max}, then Φ_x is k_{max}-derivable and one has

$$\Phi_x^{(k)}(s) \doteq \frac{d^k\Phi_x}{ds^k}(s) = \int (ix)^k e^{isx} \cdot \tilde{\mu}^x \quad , \quad k \le k_{max} \tag{4.18}$$

and hence

$$\Phi_x^{(k)}(0) = i^k m_k(x) \qquad . \tag{4.19}$$

In other words Φ_x, in the neighbourhood of $s = 0$ acts as generating function for the moments of the probability distribution:

$$\Phi_x(s) = 1 + im_1 s - \frac{1}{2} m_2 s^2 + \ldots + \frac{i^k}{k!} m_k s^k + O(s^k) \quad , \quad k \le k_{max} \tag{4.20}$$

Let us construct the characteristic function in two simple cases:-

a) x is a reduced Gaussian random variable, with probability density

$$\theta(x) = \frac{1}{\sqrt{2\pi}} e^{-\frac{1}{2}x^2} \quad . \tag{4.21}$$

One has

$$\Phi_x(s) = \frac{1}{\sqrt{2\pi}} \int_{-\infty}^{+\infty} e^{isx - \frac{1}{2}x^2} dx = e^{-\frac{1}{2}s^2} \quad . \tag{4.22}$$

On the other hand ((4.20) with $k_{max} = \infty$)

$$\Phi_x(s) = \sum_{k=0}^{\infty} \frac{i^k m_k}{k!} s^k \tag{4.23}$$

whence

$$m_k = \frac{1}{\sqrt{2\pi}} \int_{-\infty}^{+\infty} x^k e^{-\frac{1}{2}x^2} dx = \begin{cases} 0 & \text{if } k = \text{odd} \\ \dfrac{k!}{2^{\frac{k}{2}}(\frac{k}{2})!} = \dfrac{k!}{k!!} & \text{if } k = \text{even} \end{cases} \quad . \tag{4.24}$$

b) The probability distribution is Poissonian, with density c. The probability distribution for a variable x to be almost surely a positive integer, such that

$$\text{prob}\{x = m\} = e^{-c} \frac{c^m}{m!} \qquad \forall m \in \mathbb{Z}_+ \tag{4.25}$$

where $c \in \mathbb{R}$, $c > 0$, is referred to as Poisson distribution with density c and writes

$$\tilde{\mu}^x = \sum_{m=0}^{\infty} e^{-c} \frac{c^m}{m!} \delta_m \quad . \tag{4.26}$$

It is straightforward to obtain:

$$E(x) = e^{-c} \sum_{m=0}^{\infty} m \frac{c^m}{m!} = c \tag{4.27}$$

and

$$E(x^2) = e^{-c} \sum_{m=0}^{\infty} m^2 \frac{c^m}{m!} = c^2 + c \tag{4.28}$$

whence

$$\sigma^2(x) = E(x^2) - [E(x)]^2 = c \quad . \tag{4.29}$$

In other words the density is nothing but the variance of the distribution. On the other hand

$$\Phi_x(s) = \int e^{isx} \cdot \tilde{\mu}^x = e^{-c} \sum_{k=0}^{\infty} e^{isk} \frac{c^k}{k!} = e^{c(e^{is}-1)} \quad . \tag{4.30}$$

A lengthy yet straightforward calculation gives then — recalling definition (4.19) —

$$m_k = (-)^k \sum_{\ell=1}^{k} c^\ell B_{k\ell} \tag{4.31}$$

where the $B_{k\ell}$ are the Bell polynomials corresponding to the function $g(s) = e^{is} - 1$, whose explicit expression reads

$$B_{k\ell} = \sum_{\{\nu_r\}}'' \prod_{r=1}^{k} \frac{r \cdot i^{k-r\nu_r}}{\nu_r! \, (r!)^{\nu_r}} \tag{4.32}$$

and where the sum $\sum''_{\{\nu_r\}}$ is to be done over all the sets $\{\nu_r\}$ of k non-negative integers ν_r satisfying simultaneously the two conditions

$$\sum_{r=1}^{k} \nu_r = \ell \quad \text{and} \quad \sum_{r=1}^{k} r\,\nu_r = k \quad . \tag{4.33}$$

The characteristic function $\Phi_x(s)$ has interesting transformation properties with respect to homothety and translations. These are summarized in the two following formulas

$$\Phi_{\lambda x}(s) = \Phi_x(\lambda s) \quad \text{and} \tag{4.34}$$

$$\Phi_{x+a}(s) = e^{ias}\,\Phi_x(s) \tag{4.35}$$

whose demonstration is simply based on definition (4.17)

$$\Phi_{\lambda x}(s) = E(e^{i\lambda s x}) \quad \text{and} \tag{4.36}$$

$$\Phi_{x+a}(s) = E(e^{is(a+x)}) = e^{isa}E(e^{isx}) \quad . \tag{4.37}$$

A scalar random variable x is said to be Gaussian if its reduced deviation $y = \bar{x}/\sigma(x)$ is a reduced Gaussian variable. Let us set $E(x) = m$, $\sigma(x) = \sigma$. The variable $y = (x - m)/\sigma$, being reduced Gaussian has probability distribution

$$\frac{1}{\sqrt{2\pi}}\, e^{-\frac{1}{2} y^2}\, dy \quad . \tag{4.38}$$

There follows that the probability distribution for x is

$$\frac{1}{\sigma\sqrt{2\pi}} \exp\left[-\frac{1}{2\sigma^2}(x - m)^2\right] dx \quad . \tag{4.39}$$

The characteristic function of y is (Eq. (4.22))

$$\Phi_y(s) = e^{-\frac{1}{2}s^2} \quad . \tag{4.40}$$

Due to (4.34) and (4.35), the characteristic function for $x = \sigma y + m$ is therefore

$$\Phi_x(s) = e^{ims}\Phi_y(\sigma s) = e^{-\frac{1}{2}\sigma^2 s^2 + ims} \quad . \tag{4.41}$$

B) Let now $x \in \mathbb{R}^n$. Indeed let us begin with $x \in \mathbb{R}^2$, namely $x = (x^1, x^2)$, $x^\ell \in \mathbb{R}$, $\ell = 1,2$. The probability distributions for the scalar variables x^ℓ, $\ell = 1,2$ are called <u>marginal</u> distributions. They are by definition the images of $\tilde{\mu}^x$(in $\mathbb{R}^2$) under the mappings $x \to x^\ell$, $\ell = 1,2$ (i.e. the projections over the coordinate axes). If the distribution is discrete, it can be written in the form

$$\tilde{\mu}^x = \sum_{i,j} p_{ij}\, \delta_{a_i,b_j} \quad . \tag{4.42}$$

One has then

$$p_{ij} = \text{prob}(x^1 = a_i\,,\; x^2 = b_j) \quad . \tag{4.43}$$

The probability distribution for x^1 is therefore

$$\tilde{\nu}_1 = \sum_i q_i\, \delta_{a_i} \tag{4.44}$$

where

$$q_i = \text{prob}(x^1 = a_i) = \sum_j p_{ij} \quad . \tag{4.45}$$

whereas the probability distribution for x^2 is

$$\tilde{\nu}_2 = \sum_j r_j\, \delta_{b_j} \tag{4.46}$$

with

$$r_j = \text{prob}(x^2 = b_j) = \sum_i p_{ij} \quad . \tag{4.47}$$

If the distribution is continuous with density $\theta(x^1, x^2)$ the probability distributions for x^1 and x^2 will have densities

$$\zeta_\ell(x^\ell) = \int_{-\infty}^{+\infty} \theta(x^1, x^2) dx^{\gamma(\ell)} \quad , \quad \ell = 1,2 \tag{4.48}$$

where $\gamma(\ell)$ denotes the complement of ℓ, i.e. $\gamma(2) = 1$, $\gamma(1) = 2$.

Among the conditional probability distributions, we are particularly interested in the case of an event $x^1 \in A$ when $x^2 \in B$. In the discrete situation, the conditional probability distribution of x^1, for $x^2 = b_j$ is given by $\sum_i q_{i|j} \delta_{a_i}$, with $q_{i|j} = p_{ij}/r_j$. Analogously the conditional probability distribution of x^2 for $x^1 = a_i$ is given by $\sum_j r_{j|i} \delta_{b_j}$ with $r_{j|i} = p_{ij}/q_i$.

Returning to the general case $x \in \mathbb{R}^n$, one defines generalized moments as

$$m_{k_1 \ldots k_n} = E\left[(x^1)^{k_1} \ldots (x^n)^{k_n}\right] \tag{4.49}$$

which are connected in an obvious way to the characteristic function

$$\Phi_x(s_1, \ldots, s_n) = E\left[e^{i(s_1 x^2 + s_2 x^2 + \ldots + s_n x^n)}\right] \quad . \tag{4.50}$$

$\Phi_x(s_1, \ldots, s_n)$ is once more nothing but the Fourier transform of the probability distribution $\tilde{\mu}^x$ of x.

C) Let finally x be a vector random variable. One can define again the characteristic function of the probability distribution $\tilde{\mu}^x$ in the following way

$$\Phi_x(s) = E(e^{i<x,s>}) = \int e^{i<x,s>} \cdot \tilde{\mu}^x = (\mathscr{F}\tilde{\mu}^x)(s) \tag{4.51}$$

where s is now any element of the dual $\mathscr{V}^*$ of $\mathscr{V}$ or of any space $\hat{\mathscr{V}}$ in duality with $\mathscr{V}$ ($\mathscr{V}$ itself, if $\mathscr{V}$ is euclidean). Utilizing a particular basis of $\mathscr{V}$ one recovers the analytic form (4.50) obtained in the case $\mathscr{V} = \mathbb{R}^n$.

The properties (4.34), (4.35) generalize as well. Let $a \in \mathscr{V}$ and $K \in \mathscr{L}(\mathscr{V},\mathscr{W})$ (the space of continuous linear mappings of the vector space $\mathscr{V}$ in the vector space $\mathscr{W}$). One has

$$\Phi_{x+a}(s) = e^{i<a,s>} \Phi_x(s) \quad , \quad s \in \mathscr{V}' \tag{4.52}$$

$$\Phi_{K \circ x}(t) = \Phi_x(K^* \circ t) \quad , \quad t \in \mathscr{W}' \quad . \tag{4.53}$$

The proof of (4.52), (4.53) generalizes in a straightforward way (4.36) and (4.37):

$$\begin{aligned}\Phi_{x+a}(s) = E(e^{i<x+a,s>}) &= E(e^{i<a,s>} e^{i<x,s>}) = \\ &= e^{i<a,s>} E(e^{i<x,s>})\end{aligned} \tag{4.54}$$

$$\Phi_{K \circ x}(t) = E(e^{i<K \circ x,t>}) = E(e^{i<x,K^* \circ t>}) \quad . \tag{4.55}$$

5. Stochastic Independence of Random Variables

Let x^1 , x^2 be two real random variables, $\tilde{\mu}^x$ the probability distribution of the pair $x = (x^1 , x^2)$ and $\tilde{\nu}_1^{x^1} , \tilde{\nu}_2^{x^2}$ the probability distributions of x^1 and x^2 respectively. x^1 , x^2 are said to be stochastically independent if

$$\tilde{\mu} = \tilde{\nu}_1 \tilde{\nu}_2 \quad . \tag{5.1}$$

Two events A and B are independent if their respective Bernoulli variables are stochastically independent.

Recalling that the measure $\tilde{\nu}_1\tilde{\nu}_2$ is characterized by the property

$$\int f(x^1)g(x^2)\cdot\tilde{\nu}_1\tilde{\nu}_2 = \int f(x^1)\cdot\tilde{\nu}_1^{x^1}\int g(x^2)\cdot\tilde{\nu}_2^{x^2} \tag{5.2}$$

whatever the functions $f, g \in \mathscr{D}^{(0)}$; such a property is obviously true, a fortiori, for f, g square integrable. Upon defining $v(y, z) = E(yz) - E(y)E(z) = E(\bar{y}\bar{z}) = \langle\bar{y}, \bar{z}\rangle$, if (5.1) holds, (5.2) writes $v(y, z) = 0$ with $y = f(x^1)$, $z = g(x^2)$.

From there one derives the following theorem. A necessary and sufficient condition for the stochastic independence of two random variables x^1 and x^2 is that their <u>correlation function</u>

$$\rho(f(x^1), g(x^2)) \tag{5.3}$$

where $\rho(y, z) = \dfrac{\langle\bar{y}, \bar{z}\rangle}{\|\bar{y}\|\,\|\bar{z}\|} = \dfrac{v(y, z)}{\sigma(y)\,\sigma(z)}$, vanishes for arbitrary functions $f, g \in \mathscr{D}^{(0)}$ or whatever $f \in L^2(\tilde{\nu}_1)$, $g \in L^2(\tilde{\nu}_2)$. Obviously, if x^1 and x^2 are stochastically independent, so are $f(x^1)$ and $g(x^2)$. The property of complete mixing defined by Eq. (7.13) of Chapter I, is thus seen to be nothing but the statement of stochastic independence of x and $\mathscr{T}^n x$ as $n\to\infty$.

One can now summarize in the following way all the (equivalent) necessary and sufficient conditions for the stochastic independence of two events A and B:

i) prob(AB) = prob(A) prob(B)

ii) prob(A|B) = prob(A) (provided prob(B) $\neq$ 0)

iii) prob(A|B) = prob(A|C(B)) (provided prob(B) $\neq$ 0 and 1)

iv) $\rho(\eta_A, \eta_B) = 0$ [i.e. $v(\eta_A, \eta_B) = 0$].

The proof of the above statements is based on the following properties:

a) $v(\eta_A, \eta_B) = E(\eta_A, \eta_B) - E(\eta_A)\, E(\eta_B) =$

$$= \text{prob}(AB) - \text{prob}(A)\,\text{prob}(B) \tag{5.4}$$

b) $$\text{prob}(AB) = \text{prob}(A|B)\,\text{prob}(B) \tag{5.5}$$

c) $$\text{prob}(A) = \text{prob}(B)\,\text{prob}(A|B) + \text{prob}[C(B)]\,\text{prob}(A|C(B)) \quad . \tag{5.6}$$

Indeed, let us assume A and B to be independent: iv) holds by definition, i) follows from a) and ii) from b). Finally

$$v(\eta_A, \eta_{C(B)}) = v(\eta_A, 1 - \eta_B) = -\, v(\eta_A, \eta_B) = 0 \tag{5.7}$$

whence (from ii)),

$$\text{prob}(A) = \text{prob}(A|C(B)) \tag{5.8}$$

iii) follows from (5.8) and ii).

On the other hand the functions of η_A (respectively η_B) form a two-dimensional vector space whose basis is $\{\eta_A, (1-\eta_A)\}$ [η_B and $(1-\eta_B)$ respectively]. Therefore iv) implies

$$v(f(\eta_A), g(\eta_B)) = 0 \quad , \quad \forall\, f, g \tag{5.9}$$

and hence guarantees the independence of the events A and B. Moreover, iv) implies i) by a); i) implies ii) by b); and ii) implies iii) by c) and the relation $\text{prob}(B) + \text{prob}[C(B)] = 1$. Analogous definitions hold in the case of several random variables: p random variables $x^1, x^2, \ldots, x^p$ are said to be stochastically independent if the probability distribution $\tilde{\mu}^x$ of $x = (x^1, \ldots, x^p)$ is equal to the product of the probability distributions $\tilde{\nu}_1^{x^1}$ of x^1, $\tilde{\nu}_2^{x^2}$ of $x^2, \ldots, \tilde{\nu}_p^{x^p}$ of x^p:

$$\tilde{\mu}^x = \tilde{\nu}_1^{x^1} \ldots \tilde{\nu}_p^{x^p} \quad . \tag{5.10}$$

The events $A_1,\dots,A_p$ are independent if their Bernoulli variables $\eta_{A_1},\dots,\eta_{A_p}$ are pairwise stochastically independent. Notice that if (5.10) holds in the form

$$\tilde{\mu}^x = \prod_{i=1}^{p} \tilde{\nu}^{x^i} \tag{5.11}$$

then $\tilde{\nu}^{x^i}$ is necessarily the probability distribution of x^i, $i = 1,\dots,p$ and hence $x^1,\dots,x^p$ are independent. Indeed, one has for any $f \in \mathscr{D}^{(0)}(\mathbb{R})$

$$\int f(x^i) \cdot \tilde{\mu}^x = \int f(x^i) \cdot \tilde{\nu}^{x^i} \tag{5.12}$$

showing that $\tilde{\nu}^{x^i}$ is in fact that the image of $\tilde{\mu}^x$ with respect to the mapping $x \to x^i$. Moreover one can show recursively that, if $x^1,\dots,x^p$ are independent scalar random variables:

$$E(x^1,\dots,x^p) = E(x^1)\, E(x^2)\, \dots\, E(x^p) \quad . \tag{5.13}$$

There follows, based on the well-known theorem concerning the Fourier transform of a convolution product of two bounded measures,

$$\mathscr{F}(\mu \circ \nu) = \mathscr{F}(\mu)\, \mathscr{F}(\nu) \tag{5.14}$$

that given any two random variables, say x and y, stochastically independent, characterized by probability distributions $\tilde{\mu}^x$ and $\tilde{\nu}^y$ respectively, the distribution function for the variable $x+y$ is $\tilde{\mu} \circ \tilde{\nu}$ and hence its characteristic function,

$$\Phi_{x+y} = \Phi_x\, \Phi_y \quad . \tag{5.15}$$

If moreover x and y are Gaussian, then $z = x + y$ is itself Gaussian with average $E(z) = E(x) + E(y)$ and variance $\sigma^2(z) = \sigma^2(x) + \sigma^2(y)$. This follows at once from (5.15), recalling that in the present case

$$\Phi_x(s) = \exp\left[isE(x)\right] \exp\left[-\frac{1}{2} s^2 \sigma^2(x)\right] \tag{5.16}$$

$$\Phi_y(s) = \exp\left[isE(y)\right] \exp\left[-\frac{1}{2} s^2 \sigma^2(y)\right] \tag{5.17}$$

whence

$$\Phi_{x+y}(s) = \exp\left\{is\left[E(x)+E(y)\right]\right\} \exp\left\{-\frac{1}{2} s^2\left[\sigma^2(x)+\sigma^2(y)\right]\right\} . \tag{5.18}$$

Let us apply the previous notions to the calculation of the variance of a set of n random variables $\{x^1,\ldots,x^n\}$ characterized by a centered Gaussian distribution. The probability distribution is

$$\frac{\sqrt{\det \mathscr{A}}}{(2\pi)^{n/2}} \exp\left(-\frac{1}{2} x^t \mathscr{A} x\right) dx^1 \ldots dx^n . \tag{5.19}$$

It was shown in (2.7) that one can represent such a distribution in a particular basis and in (2.11) the basis B was chosen in such a way that the variables are independent reduced Gaussian variables:

$$\tilde{\mu}_B = \prod_{i=1}^{n} \frac{1}{\sqrt{2\pi}} \exp\left[-\frac{1}{2}(x_B^i)^2\right] dx_B^i \tag{5.20}$$

namely

$$\sigma^2(x^i) = 1$$

$$v(x^i , x^j) = \delta_{ij} , \quad i,j = 1,\ldots,n \tag{5.21}$$

We denote by V the matrix of elements $V_{ij} = v(x^i , x^j)$. One can recover any other basis B' by setting $x_B = Sx_{B'}$, $(x_{B'} = S^{-1} x_B)$,

$$\mathscr{A}(B') = S^t \mathbb{1}_n S = S^t S \tag{5.22}$$

and

$$V(B') = S^{-1}\,\mathbb{1}_n (S^{-1})^t = \mathscr{A}^{-1}(B') \qquad . \tag{5.23}$$

There follows the important result: if $x = (x^1,\ldots,x^n)$ has a Gaussian probability distribution and $\rho(x^i , x^j) = 0$ for $i \neq j$, the variables $x^1,\ldots,x^n$ are independent and

$$v(x^i , x^j) = \sigma_i^2\,\delta_{ij} \tag{5.24}$$

$$\tilde{\mu}^X = \prod_{i=1}^{n} \frac{1}{\sigma_i\sqrt{2\pi}} \exp\left[-\frac{1}{2}\frac{(x^i)^2}{\sigma_i^2}\right] dx^i \qquad , \tag{5.25}$$

We can now compute the characteristic function for the Gaussian distribution (5.19) of density

$$\theta(x) = \frac{\sqrt{\det \mathscr{A}}}{(2\pi)^{n/2}} \exp\left(-\frac{1}{2}\,x^t \mathscr{A} x\right) \qquad . \tag{5.26}$$

We consider (5.26) as a special representation of a probability density whose general representation is

$$\frac{\sqrt{\det \mathscr{A}(B)}}{(2\pi)^{n/2}} \exp\left[-\frac{1}{2}\,x_B^t\,\mathscr{A}(B)\,x_B\right] \qquad . \tag{5.27}$$

Let B', B'' be two basis such that $\mathscr{A}(B') = \mathscr{A}$, $\mathscr{A}(B'') = \mathbb{I}_n$. Set for simplicity, $x_{B'} = x$, $x_{B''} = x'$. The characteristic function for x' is

$$\Phi_{x'}(s') = E(e^{i\langle x', s'\rangle}) \tag{5.28}$$

where

$$\langle x', s' \rangle = \sum_k s'^k\, x'^k \tag{5.29}$$

and

$$e^{i<x',s'>} = \prod_k e^{is'^k \, x'^k} \quad . \tag{5.30}$$

Since the variables x'^k are independent, so are the variables $\exp(is'^k \, x'^k)$,

$$\Phi_{x'}(s') = \prod_k E(e^{is'^k \, x'^k}) = \prod_k \Phi_{x'^k}(s'^k) =$$

$$= \prod_k \exp\left[-\frac{1}{2}(s'^k)^2\right] = \exp\left(-\frac{1}{2} < s', s' >\right) =$$

$$= \exp\left(-\frac{1}{2} s'^t \, s'\right) \tag{5.31}$$

where use has been made of (4.40). Setting now $x' = Sx$, one finds

$$\Phi_x(s) = \Phi_{x'}\left[(S^{-1})^t s'\right] = \exp\left[-\frac{1}{2} s^t S^{-1} (S^{-1})^t s\right] \quad . \tag{5.32}$$

Due to (5.23), which in present notation writes $S^{-1}(S^{-1})^t = \mathscr{A}^{-1}$, we finally have the characteristic function corresponding to (5.26) in the form

$$\Phi_x(s) = \exp\left(-\frac{1}{2} s^t \mathscr{A}^{-1} s\right) \quad . \tag{5.33}$$

There follows, with proof quite identical to that given for the scalar case ((5.15), (5.18)), that if x, y are two independent random n-tuples of Gaussian density characterized by the variance matrices $\mathscr{A}$ and $\mathscr{B}$ respectively, the n-tuple $z = x + y$ is still Gaussian with a density

$$\frac{\sqrt{\det \mathscr{C}}}{(2\pi)^{n/2}} \exp\left(-\frac{1}{2} z^t \mathscr{C} z\right) \tag{5.34}$$

where

$$\mathscr{C}^{-1} = \mathscr{A}^{-1} + \mathscr{B}^{-1} \quad . \tag{5.35}$$

The additivity expressed by (5.35) has indeed a more general validity, connected with the following proposition. If two random variables (n-tuples) x, y are such that $\rho(x^i, y^j) = 0 \quad \forall i,j = 1,\dots,n,$ then

$$V(x^1 + y^1,\dots,x^n + y^n) = V(x^1,\dots,x^n) + V(y^1,\dots,y^n) \quad . \tag{5.36}$$

The latter identity simply follows from the observation that under hypothesis stated,

$$v(x^i + y^i, x^j + y^j) = v(x^i, x^j) + v(x^i, y^j) + v(y^i, x^j) + $$
$$+ v(y^i, y^j) = v(x^i, x^j) + v(y^i, y^j) \ . \tag{5.37}$$

(5.35) is a special case of (5.36) (see (5.23)). Let now $M \in \mathscr{L}(\mathbb{R}^n, \mathbb{R}^p)$. If x is a centered Gaussian n-tuple of density $\theta(x)$ as given by (5.26), provided M has rank p then the p-tuple $y = Mx$ is itself Gaussian centered with variance matrix

$$\mathscr{B} = (M^t)^{-1} \mathscr{A} M^{-1} \quad . \tag{5.38}$$

This follows noticing that with $\Phi_x(s)$ given by (5.33), (4.53) implies

$$\Phi_{Mx}(u) = \exp\left[-\frac{1}{2} u^t M \mathscr{A}^{-1} M^t u^t\right] \quad . \tag{5.39}$$

If, in particular, $m \in (\mathbb{R}^n)'$, the random variable $y = mx$ is centered Gaussian with variance $m \cdot \mathscr{A}^{-1} \cdot m^t$. There follows that, if x is a centered Gaussian n-tuple, all of its components are Gaussian centered.

6. Convergence of Probability Distributions and Random Variables

To begin with, the set of probability distributions has to be equipped with a topology. It is convenient to select the so-called narrow topology. Let $\mathcal{M}$ denote the set of bounded measures on $\mathbb{R}$, and L_∞ the space of continuous functions which tend to zero at infinity. One defines weak topology on $\mathcal{M}$, the weak topology in the usual sense (i.e. the topology of simple convergence on the dual space) on L_∞. A sequence of measures $\mu_n \in \mathcal{M}$ tends weakly to the measure $\mu \in \mathcal{M}$, if for all $\tilde{f} \in L_\infty$ one has

$$\lim_{n\to\infty} \int \tilde{f} \cdot \mu_n = \int \tilde{f} \cdot \mu \quad . \tag{6.1}$$

One calls narrow topology on $\mathcal{M}$, the weak topology as defined above, when $\mathcal{M}$ is considered as a subspace of the dual space of the space of bounded continuous functions. A sequence of measures $\mu_n \in \mathcal{M}$ tends narrowly to a measure $\mu \in \mathcal{M}$ if for all continuous bounded functions f one has

$$\lim_{n\to\infty} \int f \cdot \mu_n = \int f \cdot \mu \quad . \tag{6.2}$$

The narrow topology is finer than the weak topology. We say that a sequence of random variables x_n tends lawfully to the random variable x if the probability distribution $\tilde{\mu}_n$ of x_n tends narrowly to the probability distribution $\tilde{\mu}$ of x, i.e. if

$$\lim_{n\to\infty} \int f \cdot \tilde{\mu}_n = \int f \cdot \tilde{\mu} \tag{6.3}$$

for all continuous bounded functions f. A simple application of known results of the theory of Fourier transform leads then to the following extremely important theorems.

The first theorem states that:

x_n tends lawfully to x if and only if the characteristic function Φ_n of x_n simply tends to the characteristic function Φ of x. Conversely if the characteristic function Φ_n of a random variable x_n simply tends to Φ (uniformly in the neighbourhood of 0), then Φ is a characteristic function and x_n tends lawfully to a random variable x whose characteristic function is Φ.

A second theorem, known as the central limit theorem, states:

Let x_n be a sequence of scalar independent (whatever n) random variables with the same probability distribution, average m and standard deviation σ. Then the variable

$$y_n = \frac{x_1 + \ldots + x_n - nm}{\sigma\sqrt{n}} \tag{6.4}$$

tends lawfully to a reduced Gaussian variable, namely:

$$\lim_{n\to\infty} \text{Prob}\left\{\frac{1}{\sigma\sqrt{n}}\left[\sum_{k=1}^{n} x_k - m\right] \leq a\right\} = \frac{1}{\sqrt{2\pi}}\int_{-\infty}^{a} e^{-\frac{1}{2}u^2}\, du \quad . \tag{6.5}$$

In order to prove the latter theorem, one first notices that y_n is reduced. Indeed, upon considering the reduced deviation

$$x'_k = \frac{x_k - m}{\sigma} \tag{6.6}$$

of x_k, for which

$$E(x'_k) = 0 \qquad , \qquad \sigma(x'_k) = 1 \tag{6.7}$$

one can write

$$y_n = \frac{x'_1 + \ldots + x'_n}{\sqrt{n}} \quad . \tag{6.8}$$

If Φ is the common characteristic function of all the random variables x'_k, one has then

$$\Phi_{y_n}(s) = \left[\Phi\left(\frac{s}{\sqrt{n}}\right)\right]^n \quad . \tag{6.9}$$

For $n \to \infty$

$$\ell n\ \Phi_{y_n}(s) = n\ \ell n \left[\Phi\left(\frac{s}{\sqrt{n}}\right)\right] \sim n\left[\Phi\left(\frac{1}{\sqrt{n}}\right) - 1\right] \tag{6.10}$$

whence, using the property that $\Phi(0) = 1$, the right-hand side of (6.9) tends to 1 for $n \to \infty$. Now, in the neighbourhood of zero, since the variables x'_k are reduced, and by definition of characteristic function,

$$\Phi(u) = 1 - \frac{1}{2}u^2 + 0(u^3) \tag{6.11}$$

one has

$$\ell n\ \Phi_{y_n}(s) \sim n\left(-\frac{1}{2}\frac{s^2}{n}\right) = -\frac{1}{2}s^2 \tag{6.12}$$

whence

$$\lim_{n \to \infty} \Phi_{y_n}(s) = e^{-\frac{1}{2}s^2} \quad . \tag{6.13}$$

Equation (6.13) implies that y_n which tends lawfully to a variable of characteristic function $e^{-\frac{1}{2}s^2}$, — due to (4.22) — is reduced Gaussian.

The central limit theorem can be generalized in the following way: let $x_1,\ldots,x_n$ be a sequence of random, independent, centered m-tuples with the same probability distribution and whose common variance matrix is $\mathscr{A}$. Then the m-tuple $y_n = \dfrac{x_1 + \ldots + x_n}{\sqrt{n}}$ tends lawfully to a centered Gaussian m-tuple y with probability density

$$\frac{\sqrt{\det \mathscr{A}}}{(2\pi)^{n/2}} \exp\left[-\frac{1}{2}y^t \mathscr{A} y\right] \quad . \tag{6.14}$$

As for the random variables, let $\mathscr{V}$ be a vector space equipped with a probability distribution, $\mathscr{H}$ the space of random variables x such that $E(|x|^2) < \infty$. Assume in $\mathscr{H}$ the customary Hilbert space topology. We say that x_n tends to x in a quadratic average if x_n tends to x in $\mathscr{H}$, namely if

$$\lim_{n\to\infty} E(|x_n - x|^2) = 0 \quad . \tag{6.15}$$

Alternative convergence criteria, occasionally adopted instead of the previous ones are:

i) Let $v \in \mathscr{V}$, $x_n = f_n(v)$, $x = f(v)$. One says that x_n tends almost surely to x if $f_n(v)$ tends to $f(v)$ almost everywhere with respect to the probability distribution on $\mathscr{V}$.

ii) x_n is said to tend to x in probability if for all $\varepsilon > 0$,

$$\text{prob}\{|x_n - x| \geq \varepsilon\} \to 0 \quad \text{as} \quad n \to \infty \quad . \tag{6.16}$$

The following theorem is known as the <u>weak law of large numbers</u>. If $\{x_n\}$ is a sequence of independent identically distributed random variables with mean m, then for every $\varepsilon > 0$

$$\lim_{n\to\infty} \text{Prob}\left\{\left| \frac{1}{n} \sum_{k=1}^{n} x_k - m \right| \geq \varepsilon\right\} = 0 \quad . \tag{6.17}$$

It derives from the following inequality (the generalized Tchebycheff inequality): let x be a random variable and f an even non-negative function on $\mathbb{R}$, such that $f(x)$ is non-decreasing for $x \geq 0$ and $E(f(x))$ exists. Then for any $a \geq 0$

$$\text{Prob}\left\{|x| \geq a\right\} \leq \frac{E(f(x))}{f(a)} \tag{6.18}$$

(6.18) is easily proven noticing that, since f is non-negative

$$E(f(x)) = \int_{\mathbb{R}} f(x) \cdot \tilde{\mu}^X \geq \int_{-\infty}^{-a} f(x) \cdot \tilde{\mu}^X + \int_a^{\infty} f(x) \cdot \tilde{\mu}^X \quad . \tag{6.19}$$

On the other hand f is even and monotonic so that a fortiori (6.19) implies

$$E(f(x)) \geq f(a) \left[\int_{-\infty}^{-a} \tilde{\mu}^X + \int_a^{\infty} \tilde{\mu}^X \right] = f(a) \operatorname{Prob}\left\{ |x| \geq a \right\} \tag{6.20}$$

which coincides with (6.18). By choosing $f(x) = |x|^\alpha$, $\alpha \geq 0$, (6.18) writes

$$\operatorname{Prob}\left\{ |x| \geq a \right\} \leq \frac{E(|x|^\alpha)}{a^\alpha} \quad , \quad a > 0 \tag{6.21}$$

which is referred to as the Markov inequality.

A stronger result comes from the Kolmogorov inequality (which is based on (6.21) with $\alpha = 2$): let $x_1,\dots,x_n$ be independent L^2 random variables, $\sigma^2(x_k)$ the variance of x_k, and $m_k = E(x_k)$, $k = 1,\dots,n$; then for every $a > 0$

$$\operatorname{Prob}\left\{ \sup_{1 \leq k \leq n} \left| \sum_{j=1}^{k} (x_j - m_j) \right| \geq a \right\} \leq \frac{1}{a^2} \sum_{k=1}^{n} \sigma^2(x_k) \quad . \tag{6.22}$$

The theorem referred to as <u>Kolmogorov's strong law of large numbers</u> states that: if $\{x_k\}$ is a sequence of n-independent L^2 random variables, such that $\sum_{k=1}^{n} \frac{1}{k^2} \sigma^2(x_k)$ is finite, then

$$\lim_{n \to \infty} \left\{ \frac{1}{n} \sum_{k=1}^{n} [x_k - E(x_k)] \right\} = 0 \quad . \tag{6.23}$$

Finally Markov's inequality, together with the strong law of large number give the very important criterion expressed by Lyapunov's theorem.

Let $\{x_k\}$ be an independent sequence of centered random variables such that for some $\varepsilon > 0$, the absolute moments $E(|x_k|^{2+\varepsilon})$ exist and

$$\lim_{n\to\infty} \left\{ \left[\sum_{k=1}^{n} \sigma^2(x_k) \right]^{-\frac{1}{2}(2+\varepsilon)} \cdot \sum_{k=1}^{n} E(|x_k|^{2+\varepsilon}) \right\} = 0 \quad . \tag{6.24}$$

Then the central limit theorem holds for the sequence $\{x_k\}$.

Chapter 4
Elements of Random Processes Theory

1. Markov Chains

Any square matrix M whose elements M_{ij} satisfy the following properties:

$$M_{ij} \geq 0 \ , \quad \forall i,j \quad , \quad \sum_j M_{ij} = 1 \quad \forall i \tag{1.1}$$

is referred to as a stochastic matrix. One says that the evolution of a system is a Markov chain if

i) the (discrete) time instants at which the system is observed form a (denumerable) sequence $0,1,2,\ldots,m,\ldots$;

ii) at each instant the system is in a state which is an element of a finite set of states, labelled 1 to n;

iii) there exists a set, equipped with a probability distribution, whose elements are the following events — "the system is in the state i at time m";

iv) with each time m is associated a stochastic matrix $M(m)$ whose matrix element $M_{ij}(m)$ equals the conditional probability that the system can be found in the state j at time $(m+1)$ if it is in the state i at time m;

v) also the conditional probability that the system can be found in the state j at time $(m+1)$ provided it is in the state i at time m and given the whole set of states of the system at times $0,1,\ldots,(m-1)$ is equal to $M_{ij}(m)$.

The matrix $M(m)$ is called the transition matrix of the system at time m. Condition ii) is not strictly necessary and has been introduced only for the sake of simplicity. The theory generalizes in a straightforward way to the case when the state manifold is either denumerable or continuous. Condition v) expresses the Markov character of the process.

Let us introduce the n-tuple $p(m)$ defining the probability distribution of the random variable which describes the state of the system at time m:

$$p(m) = \begin{pmatrix} p^1(m) \\ \vdots \\ p^n(m) \end{pmatrix} \quad , \tag{1.2}$$

where $p^i(m)$, $i=1,\ldots,n$ is the probability of the state i at time m. Notice that

$$p^i(m) \geq 0 \quad , \quad \sum_i p^i(m) = 1 \quad \forall m \quad . \tag{1.3}$$

By the theorem of composed probabilities [Eq. (2.21) of Chapter 3], the components of $p(m+1)$ are given by

$$p^j(m+1) = \sum_i M_{ij}(m)\, p^i(m) \tag{1.4}$$

namely

$$p(m+1) = M(m)\, p(m) \quad . \tag{1.5}$$

Upon denoting by $p(0)$ the initial probability distribution of the different states, the distribution at time m is given by the recursive solution of (1.5):

$$p(m) = M(m-1)\, M(m-2) \ldots M(1)\, M(0)\, p(0) \quad m \geq 1 \quad . \tag{1.6}$$

A Markov chain is said to be <u>stationary</u> if it is characterized by a transition matrix M independent of time (m). Then (1.5) and (1.6) can be rewritten in the simpler form:

$$p(m+1) = M\, p(m) \tag{1.7}$$

whose solution is

$$p(m) = M^m\, p(0) \quad . \tag{1.8}$$

In the case of stationary Markov chains, the asymptotic behaviour of the system depends only on the initial probability distribution of states $p(0)$ and on the features of the transition matrix M. The notion which turns out to be crucial is that of allowed transitions — we say that the transition from state i to state j $(i \to j)$ is allowed if $M_{ij} \neq 0$.

In order to analyse such a notion rigorously, a few concepts from graph theory need to be introduced. Any mapping Γ of a set X over the set of its own parts, is called an <u>oriented graph</u> (shortly a graph) over X. The elements of X are called the <u>vertices</u> of the graph. Any pair of vertices (x,y) with $y \in \Gamma(x)$ is called an <u>arc</u> of the graph Γ; the element x being the origin and the element y the end-point of the arc.

If X is a finite set, one can represent its elements by points in a plane and Γ will then be represented by the set of its arcs. Figure 36 gives an example of such a representation. A sequence of vertices $x_0, x_1, \ldots, x_k$ such that $x_{i+1} \in \Gamma(x_i)$ is called a <u>walk</u>. The integer k is referred to as the length of the walk. A walk of non-zero length can be defined as well as the ordered sequence of its arcs $(x_0,x_1), (x_1,x_2), \ldots, (x_{k-1},x_k)$. The vertex x_0 is called the origin of the walk, the vertex x_k its end-point. The walk is said to be elementary if $x_j \neq x_i$ for $i \neq j$. Any walk whose origin coincides with the end-point is called a <u>circuit</u>. Every walk of zero length is a circuit. A circuit of length 1 is called a <u>loop</u>.

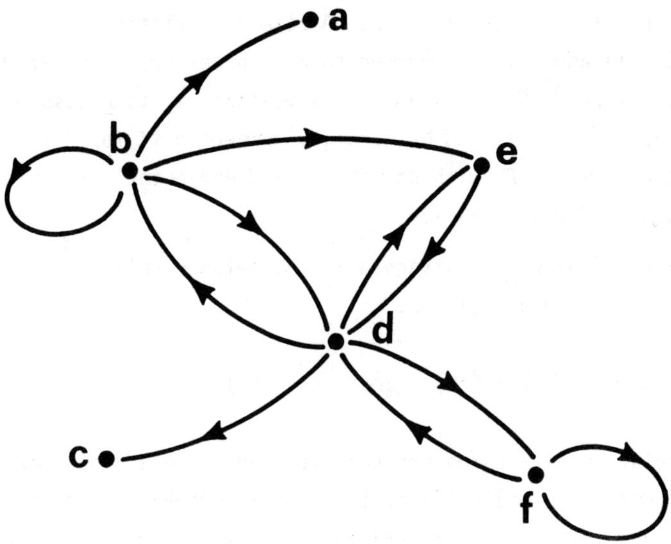

Fig. 36

Given a graph Γ, one can define the following sub-graphs

i) $y \in \Gamma^{(n)}(x)$; $n \geq 0$, if there exists a walk of length n of the graph Γ, which has x as its origin and y as its end-point. In particular $\Gamma^{(1)} = \Gamma$; $\Gamma^{(0)}(x) = \{x\}$ $\forall x \in X$.

ii) $y \in \hat{\Gamma}(x)$ if there exists a walk of any length of the graph Γ, having x as its origin and y as its end-point. Obviously

$$\hat{\Gamma}(x) = \cup \sum_{n=0}^{\infty} \Gamma^{(n)}(x) \quad . \tag{1.9}$$

Considered as an order relation on the set X, the relation $y \in \hat{\Gamma}(x)$ is a pre-order (recall that a relation here denoted $\subseteq$ is an order relation on a set Λ, if it has the following properties: i) $\lambda \subseteq \lambda$, ii) $\lambda \subseteq \mu$ and $\mu \subseteq \nu \Longrightarrow \lambda \subseteq \nu$, iii) $\lambda \subseteq \mu$ and $\mu \subseteq \lambda \Longrightarrow \lambda = \mu$,

$\lambda, \mu, \nu \in \Lambda$. A set Λ for which i), ii) and iii) hold is called an ordered set. Λ is totally ordered if the further relation holds: iv) $\lambda \subseteq \mu$ or $\mu \subseteq \lambda$, $\forall \lambda, \mu \in \Lambda$. A relation satisfying only the properties i) and ii) is referred to as a preorder). In particular $x \in \hat{\Gamma}(x)$, $\forall x \in X$. If $y \in \hat{\Gamma}(x)$, one says that y is <u>accessible</u> from x. A graph Γ is said to be strongly connected if for all $x, y \in X$, there is a walk on Γ with origin x and end-point y; in other words if $\hat{\Gamma}(x) = X$ for all $x \in X$.

With all graphs is attached an equivalence relation, called canonical, defined as follows:

$$x \sim y \quad \text{if} \quad y \in \hat{\Gamma}(x) \quad \text{and} \quad x \in \hat{\Gamma}(y) \quad . \tag{1.10}$$

Such a relation is nothing but the equivalence relation connected with the preorder. Its equivalence classes are referred to as the <u>classes</u> of the graph. The preorder relation $y \in \hat{\Gamma}(x)$ modulo the canonical equivalence relation defines an order relation over the classes of the graph. One can introduce the following terminology: a class C_2 is accessible from a class C_1 if every $x_2 \in C_2$ is accessible from all $x_1 \in C_1$ (in fact it is sufficient that this is satisfied by an arbitrary pair x_1, x_2). A subset A of X is said to be closed if

$$x \in A \ , \ y \in \hat{\Gamma}(x) \Longrightarrow y \in A \tag{1.11}$$

or, in other words, if

$$x \in A \Longrightarrow \hat{\Gamma}(x) \subset A \tag{1.12}$$

or, still differently stated, there exists no arc with its origin in A and its end-point in $C(A)$. A class is called <u>final</u> if it is a closed subset, i.e. if it is a maximal element for the order relation among classes. If there exists only a finite number of classes (in particular if X is finite), there is at least one final class. A class which is not final is said to be <u>transitory</u>.

Let now M be a stochastic matrix with n rows for an n-state system. We associate with M the graph whose set of vertices is $X \approx \{1,2,\ldots,n\}$ such that

$$j \in \Gamma(i) \quad \text{if} \quad M_{ij} > 0 \quad . \tag{1.13}$$

Interpreting M as the transition matrix of a Markov chain, the relation $j \in \Gamma(i)$ implies that the transition from the state i to j is permitted in one step. Then the matrix M^n is associated with the graph $\Gamma^{(n)}$.

2. Spectral Analysis of Stochastic Matrices

Let us introduce in $\mathbb{C}^n$ the norm

$$\|x\| = \sum_i |x^i| \tag{2.1}$$

and in $\mathscr{L}(\mathbb{C}^n, \mathbb{C}^n)$ the associated norm

$$\|A\| = \sup_{\|x\| \le 1} \|Ax\| = \sup_i \|A_i\| \tag{2.2}$$

where A_i is defined by $Ax = \sum_i x^i A_i$. Notice that one cannot have $\|Ax\| = \|A\| \, \|x\|$ unless $x^i \neq 0$ implies $\|A_i\| = \|A\|$. This follows from the fact that the second inequality in the following relation

$$\|Ax\| \le \sum_i |x^i| \; \|A_i\| \le \sup_i \|A_i\| \cdot \sum_i |x^i| \tag{2.3}$$

i.e.

$$\|Ax\| \le \|A\| \, \|x\| \tag{2.4)]}$$

cannot become an equality other than if

$$|x^i| \, \|A_i\| = |x^i| \sup_i \|A_i\| \quad . \tag{2.5}$$

Let us now consider the stochastic matrix M defined by (1.1) as an element of $\mathscr{L}(\mathbb{C}^n, \mathbb{C}^n)$. There follows from (2.2) the proposition that every stochastic matrix has norm 1. Equation (2.4) implies then that the eigenvalues of a stochastic matrix have absolute value ≤ 1. Letting $y = (1,\dots,1)$, by (1.1) we have $yM = y$, whence $M^t y^t = y^t$. In other words M^t has an eigenvalue equal to 1 and hence, so does M: every stochastic matrix has the eigenvalue 1. Let now $\mathscr{V}$ be a finite dimensional vector space, $\dim \mathscr{V} = n$. An operator $H \in \mathscr{L}(\mathscr{V}, \mathscr{V})$ is said to be nilpotent if there exists an integer k such that $H^k = 0$. Let besides $A \in \mathscr{L}(\mathscr{V}, \mathscr{V})$ be an arbitrary operator on $\mathscr{V}$ and denote by $P(s) = \det(s\mathbb{I}_n - A)$ the characteristic polynomial of A and by $s_1,\dots,s_p$ the eigenvalues of A, of multiplicities $k_1,\dots,k_p$ respectively and solutions of the secular equation

$$P(s) = 0 \quad . \tag{2.6}$$

We have then the following spectral decomposition theorem which is stated in two equivalent forms:

a) There exists a decomposition of $\mathscr{V}$ as a direct sum of subspaces $\mathscr{V}_1,\dots,\mathscr{V}_p$ satisfying the following conditions:

   i) $\mathscr{V}_1,\dots,\mathscr{V}_p$ are stable with respect to A,

   ii) $\dim \mathscr{V}_i = k_i$, $i = 1,\dots,p$,

   iii) the restriction of A to $\mathscr{V}_i$, say $A_{\mathscr{V}_i}$, has eigenvalue s_i, namely it has the form of $s_i \mathbb{I}_{k_i} + \tilde{H}_i$, with $\tilde{H}_i$ nilpotent $(\tilde{H}_i^{k_i} = 0)$, and

   iv) $\mathscr{V}_i = \ker\left((s_i \mathbb{I}_n - A)^{k_i}\right)$.

b) There exist projectors Π_i and nilpotent operators H_i, $i = 1,\dots,p$ such that

   i) $\sum_i \Pi_i = \mathbb{I}_n \quad , \quad \Pi_i\Pi_j = 0$ for $i \neq j$, $\Pi_i^2 = \Pi_i$

   ii) the set of $y_i = \pi_i(v)$, $v \in \mathscr{V}$ is stable with respect to

A, $i=1,\dots,p$,

iii) $H_i\Pi_i = H_i$,

iv) $A = \sum_i (s_i \mathbb{I}_{k_i} + H_i)\Pi_i$.

The connection between the two statements of the theorem is obtained by noticing the following identifications:

i) the set of y_i with $\mathscr{V}_i$ and

ii) the restriction of H_i to $\mathscr{V}_i$ with $\tilde{H}_i$.

The subspace $\mathscr{V}_i$ is referred to as the spectral subspace and the projector Π_i as the spectral projector associated with the eigenvalue s_i.

We can now state the three following lemmas.

I) If H is nilpotent, $\ker H^{k-1}$ is strictly contained in $\ker H^k$, whatever k is, provided H^{k-1} is different from the whole space. The proof is simple as one has, trivially,

$$\ker(H^k) \supset \ker(H^{k-1}) \quad . \tag{2.7}$$

On the other hand, suppose $\ker(H^k) = \ker(H^{k-1})$; then we should have

$$\ker(H^{k+1}) = H^{-1}(\ker(H^k)) = H^{-1}(\ker(H^{k-1})) = \ker(H^k) \tag{2.8}$$

whence

$$\ker(H^p) = \ker(H^{k-1}) \quad \text{for} \quad p \geq k-1 \quad . \tag{2.9}$$

The latter identity is however absurd if $\ker(H^{k-1})$ does not coincide with the whole space because for p large enough, $H^p = 0$.

II) Let $\mathscr{L}(\mathscr{V},\mathscr{V})$ be equipped with the norm defined in (2.2), associated with the norm $\|\cdot\|$ in $\mathscr{V}$. If H is nilpotent (non zero) and $z \in \mathbb{C}$, one has

$$\|z\,\mathrm{I}_n + H\| > |z| \quad . \tag{2.10}$$

Equation (2.10) is proved noticing first that $\|z\,\mathrm{I}_n + H\| \geq |z|$ since $z\,\mathrm{I}_n + H$ has z as its only eigenvalue. Assuming with no loss of generality $z = 1$, consider then $\mathrm{I}_n + H$. Let $v \in \mathscr{V}$ be such that $Hv \neq 0$ whereas $H^2 v = 0$. One has

$$(\mathrm{I}_n + H)^k v = (\mathrm{I}_n + kH)v \tag{2.11}$$

whence,

$$\lim_{k\to\infty} \|(\mathrm{I}_n + H)^k v\| = \infty \quad . \tag{2.12}$$

On the other hand, by (2.4) and (2.11),

$$\|(\mathrm{I}_n + H)^k v\| = \|(\mathrm{I}_n + kH)v\| \leq \|\mathrm{I}_n + kH\|\ \|v\| \leq$$

$$\leq \|\mathrm{I}_n + H\|^k\ \|v\| \quad . \tag{2.13}$$

By (2.12), (2.13) then

$$\|\mathrm{I}_n + H\| > 1 \quad . \tag{2.14}$$

III) Let $A \in \mathscr{L}(\mathscr{V},\mathscr{V})$ and z an eigenvalue of A such that $|z| = \|A\|$. Then the nilpotent associated with z is equal to zero.

The proof is as follows. Let M be the spectral subspace associated with z. The restriction of A to M has the form $z\,\mathrm{I}_n + H$. If $H \neq 0$, then by (2.10) $\|A\| \geq \|z\,\mathrm{I}_n + H\| > |z|$, contrary to the hypothesis.

The three lemmas I, II, III) imply the following proposition. For any stochastic matrix M, the nilpotents associated with the eigenvalues of absolute value 1 are zero. In other words, if z is an eigenvalue of modulus 1 and multiplicity k, the vector subspace of the eigenvectors associated with it has dimension k.

We define support of an element $x \in \mathbb{C}^n$ the set of indices i such that $x^i \neq 0$. Let α be a set of indices. Denote by $\mathcal{E}_\alpha$ the set of elements with support in α, namely the set of elements x such that $x^i = 0$ for $i \notin \alpha$. If α is not empty, a necessary and sufficient condition for it to be closed is that $\mathcal{E}_\alpha$ is stable with respect to M, namely that M is partitioned in sub-matrices in the following way:

$$M = \left|\begin{array}{c|c} M^\alpha_\alpha & M^\alpha_\beta \\ \hline 0 & M^\beta_\beta \end{array}\right| \quad . \tag{2.15}$$

Now, if x is the eigenvector of a stochastic matrix M associated with an eigenvalue of absolute value 1, the support of x is a closed set. Indeed, let

$$x = \left|\begin{array}{c} x^\alpha \\ \hline 0 \end{array}\right| \quad \text{with} \quad \begin{cases} x^i \neq 0 & \text{if } i \in \alpha \\ Mx = \lambda x \ , & |\lambda| = 1 \end{cases} \tag{2.16}$$

and

$$M = \left|\begin{array}{c|c} M^\alpha_\alpha & M^\alpha_\beta \\ \hline M^\beta_\alpha & M^\beta_\beta \end{array}\right| \quad . \tag{2.17}$$

One has

$$M^\alpha_\alpha \, x^\alpha = \lambda x^\alpha \tag{2.18}$$

whence

$$\|M_{\alpha}^{\alpha}\| \, \|x^{\alpha}\| = \|x^{\alpha}\| \tag{2.19}$$

that implies $\|M_{\alpha}^{\alpha}\| = 1$. There follows by (2.5) that all the columns of M_{α}^{α} have norm equal to 1 and hence $M_{\alpha}^{\beta} = 0$, by definition of stochastic matrix. Then, according to (2.15), α is closed.

We say that a stochastic matrix is <u>irreducible</u> if $\mathcal{E}_{\alpha}$ is stable only for $\alpha = \emptyset$ and $\alpha = \{1,2,\ldots,n\}$, i.e. if M has no closed sets other than its empty or full parts (there exist no partition of M of the form (2.17)). M is irreducible if and only if there exists a single class, i.e. if any state j is accessible from every state i. In fact, if M is irreducible, any final class has to be a full part of it. The full part is therefore a class and there exists one class only. The converse is immediate.

We have the following property: the eigenvalues of modulus 1 of an irreducible stochastic matrix M are simple. In order to prove it, let us observe that if s is a multiple eigenvalue of absolute value 1, the subspace of associated eigenvectors has dimension ≥ 2. It contains therefore non-vanishing vectors having at least one component equal to zero.

In other words there exist eigenvalues whose support is different from both the empty and the full part. There follows the existence of a non-trivial closed set and M is not irreducible which contradicts the assumption. Let $\alpha, \beta, \ldots, \lambda$ denote the final classes of M. We have the following partition:

$$M = \left(\begin{array}{cccc|c} M_{\alpha}^{\alpha} & & 0 & & \\ \hline & M_{\beta}^{\beta} & & & \\ 0 & & \ddots & & R \\ & & & M_{\lambda}^{\lambda} & \\ \hline & & 0 & & Q \end{array}\right) \tag{2.20}$$

where R denotes any non-zero submatrix. We want to show that Q has no eigenvalue of modulus 1, and therefore the only eigenvalues of absolute value 1 of M are those of the stochastic matrices $M^{\alpha}_{\alpha}, M^{\beta}_{\beta},\ldots,M^{\lambda}_{\lambda}$ induced on the final classes. Equation (2.20) implies the following partition:

$$M^m = \left(\begin{array}{cccc|c} (M^m)^{\alpha}_{\alpha} & & & & \\ & (M^m)^{\beta}_{\beta} & & & \\ & & \ddots & & S \\ & & & (M^m)^{\lambda}_{\lambda} & \\ \hline & & & & Q^m \end{array}\right) \qquad (2.21)$$

where $(M^m)^{\alpha}_{\alpha} = (M^{\alpha}_{\alpha})^m,\ldots,(M^m)^{\lambda}_{\lambda} = (M^{\lambda}_{\lambda})^m$. Let $\mathscr{A} = \alpha \cup \beta \cup \ldots \cup \lambda$ be the union of the final classes and $\mathscr{B} = C(\mathscr{A})$. For every $i \in \mathscr{B}$ there exists a walk of origin i and end-point j, $j \in \mathscr{A}$. Let ℓ_i denote the length of such a walk. One can add to it any walk contained in $\mathscr{A}$ of arbitrary length; therefore for any $\ell \geq \ell_i$ there exists a walk of length ℓ starting from i and landing in $\mathscr{A}$.

Finally let us choose $m = \sup_{i \in \mathscr{B}} \ell_i$, and consider M^m. By construction there exists in every column of S a non-vanishing element. There follows that $\|Q^m\| < 1$, and hence Q^m cannot have any eigenvalue of absolute value 1. The same of course holds for Q. A corollary of the above proposition is that the multiplicity of the eigenvalue 1 equals the number of final classes. Thus the search of eigenvalues of absolute value 1 is limited to the case of irreducible stochastic matrices. We will therefore expand now the study of the combinatorial properties of such matrices.

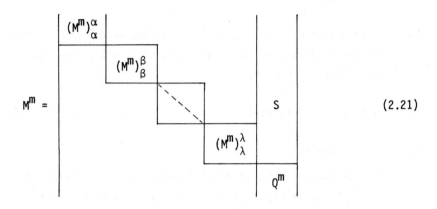

One defines <u>period</u> of an irreducible stochastic matrix M the greatest common divisor (g.c.d.) of the lengths of its circuits. The first interesting proposition is as follows. Let M be an irreducible stochastic matrix of period p. Let ℓ_1, ℓ_2 be the lengths of two walks with the same origin and end-point. One has

$$\ell_1 \equiv \ell_2 \pmod p \quad . \tag{2.22}$$

This can be proved by noting that if i is the origin and j the end-point and we denote by ℓ_3 the length of a walk going from j to i, by definition of period,

$$\ell_1 + \ell_3 \equiv \ell_2 + \ell_3 \pmod p \tag{2.23}$$

whence (2.22) follows.

The equivalence class of the following equivalence relation: "the length of any walk going from i to j is a multiple of p" is referred to as the subclass of an irreducible stochastic matrix M. We have then p subclasses denoted $c_0, c_1, \ldots, c_{p-1}$ such that any walk with origin in c_h and ending in c_k has length congruent to (k - h) (mod p). In fact, let's consider an arbitrary state v, and let ℓ_1, ℓ_2 be the lengths of two walks of origin v and end-points i and j respectively. Let moreover ℓ_3 be the length of a walk with origin i and end-point j. One has

$$\ell_1 + \ell_3 \equiv \ell_2 \pmod p \quad . \tag{2.24}$$

If i and j belong to the same subclass, then $\ell_3 \equiv 0 \pmod p$ and hence

$$\ell_1 \equiv \ell_2 \pmod p \quad . \tag{2.25}$$

Conversely if $\ell_1 \equiv \ell_2 \pmod p$ one has $\ell_3 \equiv 0 \pmod p$, and i and j belong to the same subclass. In other words every subclass is characterized in a one-to-one way by the length, modulo p, of the walks going from v to a point of such a subclass. Let now c_h, $h=0,1,\ldots,(p-1)$ be the subclass of the states i, such that one goes from v to i by a walk of length congruent to h. Choose $i \in c_h$, $j \in c_k$, and define once more: ℓ_1 the length of the walk going from v to i, ℓ_2 the length of the walk going from v to j, ℓ_3 the length of the walk going from i to j. One has

$$\begin{aligned} &\ell_1 \equiv h \pmod{p} \\ &\ell_2 \equiv k \pmod{p} \qquad (2.26) \\ &\ell_1 + \ell_3 \equiv \ell_2 \pmod{p} \end{aligned}$$

whence

$$\ell_3 = k - h \pmod{p} \qquad (2.27)$$

which completes the demonstration.

Introducing the convenient notation $c_r = c_q$ if $r \equiv q \pmod{p}$, the following corollary follows from (2.27). Let $i_0, i_1, \ldots, i_k$ be the successive steps (states) of a walk. If $i_0 \in c_r$, then $i_k \in c_{r+k}$. In particular, if $j \in \Gamma(i)$ and $i \in c_r$, then $j \in c_{r+1}$; if $j \in \Gamma^{(p)}(i)$, i and j then belong to the same subclass. Denoting $M_h^k \doteqdot M_{c_h}^{c_k}$, the following partition of M corresponds to the partition of the sets of states into subclasses:

$M =$

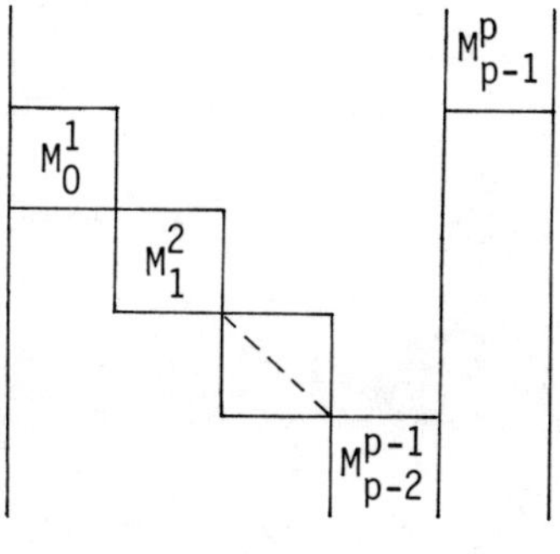

(2.28)

Since the subclasses do not necessarily have the same number of states, the matrices M_h^k are not necessarily square. The following proposition due to Frobenius can then be derived: an irreducible stochastic matrix of period p has as (simple) eigenvalues of absolute value 1, the roots of unit solutions of the equation $s^p = 1$. The eigenvectors associated to the eigenvalue 1 have all the phases of their components equal. In fact, let s and $x \neq 0$ be such that

$$Mx = sx \quad , \quad |s| = 1 \quad . \tag{2.29}$$

Recall that $x^i \neq 0$, whatever i is. The first relation in (2.29) writes

$$\sum_i M_i^j \, x^i = sx^j \tag{2.30}$$

with

$$\sum_j \Big| \sum_i M_i^j \, x^i \Big| = \sum_i |x^i| \tag{2.31}$$

where the second relation as well was used. Since by definition

$$\sum_j M_i^j = 1 \qquad \forall i \tag{2.32}$$

one can write,

$$\sum_i |x^i| = \sum_i \sum_j M_i^j \, |x^i| \tag{2.33}$$

whereby (2.31) reads

$$\sum_j \Big| \sum_i M_i^j \, x^i \Big| = \sum_j \sum_i M_i^j \, |x^i| \quad . \tag{2.34}$$

On the other hand, obviously

$$| \sum_i M_i^j x^i | \leq \sum_i M_i^j |x^i| \tag{2.35}$$

so that one can deduce

$$| \sum_i M_i^j x^i | = \sum_i M_i^j |x^i| = \sum_i | M_i^j x^i | \quad , \quad \forall j \quad . \tag{2.36}$$

There follows that, for all j, $\mathrm{Arg}(M_i^j x^i)$ is independent of i (for all i such that $M_i^j \neq 0$). Due to (2.30) this can be written

$$\mathrm{Arg}(M_i^j x^i) = \mathrm{Arg}(sx^j) \tag{2.37}$$

(for all i and j such that $M_i^j \neq 0$), whence

$$\text{for } j \in \Gamma(i) \quad , \quad \mathrm{Arg}(x^i) = \mathrm{Arg}(sx^j) \quad . \tag{2.38}$$

One can deduce that for all circuits of length ℓ, one has $s^\ell = 1$. On the other hand, the set of integers m, such that $s^m = 1$ is an ideal, which contains all cycle lengths. In particular it contains the greatest common divisor of the cycle lengths, p, that implies

$$s^p = 1 \quad . \tag{2.39}$$

Let us remark that if $s = 1$ and $j \in \Gamma(i)$, (2.38) reads

$$\mathrm{Arg}(x^i) = \mathrm{Arg}(x^j) \quad . \tag{2.40}$$

Now the condition $j \in \Gamma(i)$ is redundant, in that any two states are connected by a walk.

Finally, upon setting $x \equiv (x^0, x^1, \ldots, x^{p-1})$, with M defined by (2.28), one finds (indices mod p):

$$Mx = (M^0_{p-1}\ x^{p-1}\ ,\ M^1_0\ x^0\ ,\ M^2_1\ x^1\ ,\ldots,\ M^{p-1}_{p-2}\ x^{p-2}) \quad . \tag{2.41}$$

On the other hand, M^p is block-diagonal with blocks $(M^p)^j_j$, $j = 0,1,\ldots,p-1$. In particular,

$$(M^p)^0_0 = M^0_{p-1} \ldots M^2_1\ M^1_0 \tag{2.42}$$

and in general

$$(M^p)^j_j = M^j_{j-1} \ldots M^1_0\ M^0_{p-1} \ldots M^{j+2}_{j+1}\ M^{j+1}_j \quad . \tag{2.43}$$

The eigenvalue equation $Mx = sx$ reads then

$$\begin{aligned} M^0_{p-1}\ x^{p-1} &= sx^0 \quad , \\ M^1_0\ x^0 &= sx^1 \quad , \\ M^2_1\ x^1 &= sx^2 \quad , \\ &\vdots \\ M^{p-1}_{p-2}\ x^{p-2} &= sx^{p-1} \quad . \end{aligned} \tag{2.44}$$

Choose $x^0 \neq 0$ such that $(M^p)^0_0 x^0 = x^0$. Upon multiplying the last (p-1) equations in (2.44) one gets

$$M^{p-1}_{p-2} \ldots M^2_1\ M^1_0\ x^0 = s^{p-1}\ x^{p-1} \tag{2.45}$$

and further multiplying both sides by M^0_{p-1} and recalling (2.42)

$$(M^p)^0_0\ x^0 = s^{p-1}\ M^0_{p-1}\ x^{p-1} \quad . \tag{2.46}$$

Since $s^p = 1$, (2.46) reads

$$sx^0 = M^0_{p-1}\, x^{p-1} \tag{2.47}$$

which coincides with the first equation in (2.44), thus showing that x is indeed the eigenvector associated with the eigenvalue s. Let us say that the matrix M^p admits p final classes, each having a single period equal to 1.

We can summarize all previous results in the following theorem. Let M be a stochastic matrix with p final classes of periods p_j respectively $(j = 0,\dots,p-1)$. The eigenvalues s of M of absolute value 1 are:

1) $s = 1$, of mutliplicity p

2) the roots of unit, solutions of $s^{p_j} = 1$, different from 1, whose multiplicity is of course the number of equations of this type of which they are solutions.

A stochastic matrix M is now defined to be <u>ergodic</u> if the eigenvalue 1 is simple, namely if there exists just one final class. A stochastic matrix is said to be <u>regular</u> if the eigenvalue 1 is simple and the period of the corresponding final class is 1. M is said to be <u>primitive</u> if it does not admit eigenvalues of absolute value 1 other than 1, namely all of its final classes have period 1.

3. Asymptotics

If $|s| < 1$ and H is nilpotent, $(H^{k+1} = 0)$, then $(s\,\mathbb{I}_n + H)^m$ tends to zero for $m \to \infty$. In fact we have

$$(s\,\mathbb{I}_n + H)^m = s^m\,\mathbb{I}_n + ms^{m-1}H + \dots + m(m-1)\,\dots\,(m-k+1)s^{m-k}H^k \tag{3.1}$$

and each factor on the right-hand side tends separately to zero as $m \to \infty$. Let then

$$M = \sum_r (s_r \mathbb{I}_n + H_r)\Pi_r \tag{3.2}$$

be the spectral decomposition of M with $|s_r| \le 1$, and $H_r = 0$ if $|s_r| = 1$; where $\{s_r\}$ denote the eigenvalues of M, Π_r the spectral projectors and the H_r's are nilpotent. We have then

$$M^m = \sum_r \Pi_r(s_r \mathbb{I}_n + H_r)^m = \sum_{|s_r|=1} \Pi_r s_r^m + \varepsilon(m) \tag{3.3}$$

where $\varepsilon(m)$ is such that $\lim_{m\to\infty} \varepsilon(m) = 0$.

This allows us to show that the following three statements are equivalent:

i) M is primitive

ii) M^m has a limit for $m \to \infty$

iii) whatever $x(0)$, $x(m) = M^m x(0)$ has a limit for $m \to \infty$.

Moreover if the above properties are verified,

$$\lim_{m\to\infty} M^m = \Pi_1 \tag{3.4}$$

where Π_1 is the projector associated with the eigenvalue 1. The proof proceeds by elementary checks:

i) $\Rightarrow$ ii) : if M is primitive, then by definition $\lim_{m\to\infty} M^m = \Pi_1$.

ii) $\Rightarrow$ iii) : trivially.

ii) $\Rightarrow$ i) : if M^m has a limit, so has $\Pi_r M^m = \Pi_r s_r^m$ and hence s_r^m, thus either $s_r = 1$ or $|s_r| = 1$.

iii) $\Rightarrow$ ii) : if $M^m x(0)$ has a limit, whatever $x(0)$, then $(M^m)_i$ has a limit and so has M^m.

Now M is regular if and only if M is primitive and Π_1 has rank 1. There follows the necessary and sufficient condition for the existence of $x_\infty = \lim_{m\to\infty} x(m)$, whatever $x(0)$, (indeed for x_∞ to be independent of $x(0)$) namely that M is regular.

Let us consider now the matrix

$$P(m) = \frac{1}{m}\left(\mathbb{I}_n + M + \ldots + M^{m-1}\right) \quad . \tag{3.5}$$

$P(m)$ is the average of m stochastic matrices and is therefore stochastic itself. For $m\to\infty$, recalling (3.3) one can write

$$P(m) = \sum_{|s_r|=1} \Pi_r \left(\frac{1}{m}\sum_{k=0}^{m-1} s_r^k\right) + \bar{\varepsilon}(m) \tag{3.6}$$

where

$$\bar{\varepsilon}(m) = \frac{1}{m}\sum_{k=0}^{m-1} \varepsilon(k) \quad . \tag{3.7}$$

Since $\varepsilon(k)\to 0$ for $k\to\infty$, $\bar{\varepsilon}(m)\to 0$ for $m\to\infty$. On the other hand

$$\lim_{m\to\infty}\left(\frac{1}{m}\sum_{k=0}^{m-1} s_r^k\right) = \begin{cases} \lim_{m\to\infty}\left(\frac{1}{m}\frac{s_r^m - 1}{s_r - 1}\right) = 0 & \text{if} \quad s_r \neq 1 \\ 1 & \text{if} \quad s_r = 1 \end{cases} \tag{3.8}$$

whence

$$\lim_{m\to\infty} P(m) = \Pi_1 \quad . \tag{3.9}$$

In particular, (3.9) implies

$$\lim_{m\to\infty} \frac{1}{m}\left[x(0) + \ldots + x(m-1)\right] = \Pi_1 x(0) \quad . \tag{3.10}$$

When property (3.10) holds, one says that the sequence $x(0),\dots,x(m-1)$ has a limit in the sense of Fejer. The necessary and sufficient condition for the existence of such a limit (and for the limit to be independent of $x(0)$) is that Π_1 is of rank 1, namely that M is ergodic. Notice that if M is ergodic, not only Π_1 has rank 1 but all of its columns are equal, since Π_1 is stochastic. The common value of the columns is the only vector w invariant with respect to M and satisfying the conditions $\sum_i w^i = 1$ and $w^i \geq 0$. w is the limit in the sense of Fejer of $x(m)$, whatever $x(0)$. If M is regular, w is the limit of $x(m)$ (whatever $x(0)$). Returning to the interpretation of M as the transition matrix of a Markov chain, one can say that w is the limit probability distribution of the states of the system.

4. Stationary Random Processes of Second Order

Consider now a random process represented by a dynamical variable x whose value at each time t is a random variable $x(t) \in \mathscr{V}$, for a vector space $\mathscr{V}$ equipped with a probability distribution. Let $\mathscr{H}$ be the space of random variables z such that $E(|z|^2)$ is finite. A random process consisting of a map $X : t \to x(t)$ which associates a (scalar or vector) random variable $x(t)$ with all times $t \in \mathbb{R}$ is said to be <u>centered</u> if, for any t, $E(x(t)) = 0$. Such a process is moreover said <u>stationary</u> if for any arbitrary set of times $t_1,\dots,t_k$, the probability distribution of the variables $x(t_1),\dots,x(t_k)$ is identical with the probability distribution of $x(t_1 - \theta),\dots,x(t_k - \theta)$ for any $\theta \in \mathbb{R}$. When $x(t)$ is a scalar, the process is called <u>stationary of the second</u> order if

$$\begin{cases} x(t) \in \mathscr{H} \quad \forall\, t \in \mathbb{R} \\ E(x(t)) \text{ is independent of } t \\ v(x(t_1 - \theta),\, x(t_2 - \theta)) = v(x(t_1),\, x(t_2)) \quad \forall\, t_1, t_2, \theta \in \mathbb{R} \end{cases} \tag{4.1}$$

Notice that if a process X is stationary, it is stationary of the second order provided $x(t) \in \mathcal{H}$. Given a process X stationary of the second order, the process $\bar{X}$,

$$\bar{X} : t \to \bar{x}(t) = x(t) - m_x$$

$$m_x = E(x(t)) \tag{4.2}$$

is stationary of the second order and centered. The constant m_x is referred to as the average of the process.

X is said to be <u>continuous</u> in quadratic average if the map $t \to x(t)$ is a continuous map of $\mathbb{R}$ in $\mathcal{H}$. All the above concepts can be generalized to the case when $x(t)$ is a multiple $(x^1(t),\ldots,x^m(t))$; m_x is then itself a multiple. In such a case, if $t \to x(t)$ is stationary of the second order, so is any process $X^\alpha : t \to x^\alpha(t)$ where α is a set of indices contained in $\{1,\ldots,m\}$, and so is — in particular — any process $X^i : t \to x^i(t)$. On the contrary, the fact that each process X^i is stationary of the second order does not necessarily imply that X is stationary of the second order. In order that this happens for a continuous centered process X, the necessary and sufficient condition is that there exists a unitary continuous representation R of $\mathbb{R}$ in $\mathcal{H}$, $R : \theta \to U_\theta$, $\theta \in \mathbb{R}$, such that

$$x^i(t - \theta) = U_\theta \, x^i(t) \quad . \tag{4.3}$$

Recall that one defines as continuous unitary representation of $\mathbb{R}$ in the Hilbert space $\mathcal{H}$ a map $\theta \to U_\theta$ of $\mathbb{R}$ into the set of unitary operators on $\mathcal{H}$ satisfying the following properties:

i) $U_{\theta+\sigma} = U_\theta \, U_\sigma$

ii) $U_{-\sigma} = (U_\sigma)^{-1} = U_\sigma^\dagger$ $\qquad\qquad\qquad$ (4.4)

iii) $\forall h \in \mathcal{H}$ the map $\theta \to U_\theta h$ is a continuous map of $\mathbb{R}$ in $\mathcal{H}$ (equipped with the strong topology defined by the norm).

[Notice that from (4.4) i) and ii) one gets:

$$U_0 = U_{\sigma-\sigma} = U_\sigma U_{-\sigma} = U_\sigma (U_\sigma)^{-1} = \mathbb{1} \quad . \qquad (4.5)]$$

The proof of (4.3) proceeds as follows. Necessary condition: if the representation R exists, one has

$$v(x^i(t_1 - \theta),\, x^j(t_2 - \theta)) = \langle x^i(t_1 - \theta),\, x^j(t_2 - \theta) \rangle =$$

$$= \langle U_\theta\, x^i(t_1),\, U_\theta\, x^j(t_2) \rangle = \langle x^i(t_1),\, x^j(t_2) \rangle =$$

$$= v(x^i(t_1),\, x^j(t_2)) \quad . \qquad (4.6)$$

Moreover the map $t \to x^i(t) = U_{-t}\, x^i(0)$ is manifestly continuous. Sufficient condition: let us consider the subspace $\mathscr{N}_x$ of $\mathscr{H}$, generated by the elements of the form

$$\xi = \sum_{i,h} a_i^h\, x^i(t_h) \qquad (4.7)$$

for an arbitrary choice of the set of discrete times $\{t_h\}$. Let us moreover consider the set of elements

$$\eta = \sum_{i,h} a_i^h\, x^i(t_h - \theta) \qquad (4.8)$$

and define the set Γ of pairs $\binom{\xi}{\eta}$. Upon denoting by $\mathscr{H}_x$ the closed subspace of $\mathscr{H}$ generated by $x^i(t)$ for $t \in \mathbb{R}$, $i = 1,\dots,m$. $\mathscr{H}_x$ is called the adherence of $\mathscr{N}_x$. Γ is a vector subspace of $\mathscr{H}_x X \mathscr{H}_x$. In general Γ is not closed. Let $\binom{\xi'}{\eta'} \in \Gamma$ be a different element of Γ, ξ' and η' being defined as in (4.7), (4.8) with a different set of coefficients, say $\{a_i'^h\}$. One has then

$$< \eta,\eta' > = \sum_{ih,jk} a_i^{h*} \, a_j^{!k} < x^i(t_h - \theta), \, x^j(t_k - \theta) > =$$

$$= \sum_{ih,jk} a_i^{h*} \, a_j^{!k} < x^i(t_h), \, x^j(t_k) > = < \xi,\xi' > \tag{4.9}$$

whence, in particular

$$\|\xi\| = \|\eta\| \tag{4.10}$$

and, Γ being a vector subspace,

$$\|\xi - \xi'\| = \|\eta - \eta'\| \quad . \tag{4.11}$$

There follows that $\xi = \xi'$ implies $\eta = \eta'$.

In other words, for any $\xi \in \mathscr{N}_X$ there exists one and only one $\eta \in \mathscr{N}_X$ such that $\binom{\xi}{\eta} \in \Gamma$. Let us set

$$\eta = U_\theta \xi \quad . \tag{4.12}$$

U_θ is a linear operator defined on the dense subspace $\mathscr{N}_X$ of $\mathscr{H}_X$, satisfying

$$< U_\theta \xi, \, U_\theta \xi' > = < \xi, \, \xi' > \tag{4.13}$$

which by continuity can be prolonged into an isometric operator on $\mathscr{H}_X$ (still denoted by U_θ, since no ambiguity arises) such that

$$U_\theta x^i(t) = x^i(t - \theta) \qquad \forall \, i \quad . \tag{4.14}$$

Condition (4.14) defines U_θ entirely on $\mathscr{H}_X$; whereby one can deduce, as required by (4.4),

$$U_{\theta_1 + \theta_2} = U_{\theta_1} \, U_{\theta_2}$$

$$U_{-\theta} = (U_\theta)^{-1} \quad . \tag{4.15}$$

Then the mapping $\theta \to U_\theta$ is a unitary representation of $\mathbb{R}$ in $\mathscr{H}_X$. Such a representation is continuous, in fact if $\theta_n \to \theta$ one has $U_{\theta_n}\xi \to U_\theta \xi$ for all $\xi \in \mathscr{N}_X$. Since $\|U_{\theta_n}\| = 1$, the topology of the simple convergence on $\mathscr{N}_X$ coincides with that on $\mathscr{H}_X$; hence $U_{\theta_n} x \to U_\theta \xi$ for all $x \in \mathscr{H}_X$ as well. One can finally obtain a prolongation of U_θ on the whole $\mathscr{H}$, by setting $U_\theta \xi = \xi$ for $\xi \in \mathscr{H} \backslash \mathscr{H}_X$.

The previous theorem can be easily generalized to the case when X is not centered, by adding to (4.3) the condition

$$U_\theta 1 = 1 \qquad . \tag{4.16}$$

Indeed the condition is sufficient because — restricting one's attention, with no loss of generality, to the case when $\mathscr{H} \equiv \mathscr{H}_X$ — the operator U_θ obviously permutes with the orthogonal projection on $\bar{\mathscr{H}} = \mathscr{H}_{\bar{X}}$, since $\{1\}$ (by (4.16)) and $\mathscr{H}_{\bar{X}}$ (by (4.1)) are stable under U_θ, and hence

$$\bar{x}^i(t-\theta) = U_\theta\, \bar{x}^i(t) \tag{4.17}$$

namely the process $\bar{x}$ is continuous stationary of the second order. On the other hand

$$E(x(t)) = <1, x(t)> = <1, U_t^\dagger\, x(0)> =$$

$$= <U_t 1, x(0)> = <1, x(0)> E(x(0)) \qquad . \tag{4.18}$$

The condition is sufficient as well. In fact, all one has to do is to select a U_θ such that

$$U_\theta \bar{x}(t-\theta) = U_\theta \bar{x}(t)$$

$$U_\theta \xi = \xi \qquad , \qquad \xi \in \mathscr{H} \backslash \mathscr{H}_X \qquad . \tag{4.19}$$

We shall henceforth consider only continuous random processes. Let $Z : t \to \begin{pmatrix} x(t) \\ y(t) \end{pmatrix}$ be a centered stationary random process of the second kind, $x(t)$, $y(t)$ being scalar random variables. The function

$$c(\theta) = v(x(t-\theta), y(t)) = \langle x(t-\theta), y(t) \rangle$$

$$= \langle U_\theta\, x(t), y(t) \rangle \tag{4.20}$$

is referred to as the correlation function of x and y. It is often denoted as $\phi_{x(t),y(t)}$ or rather, since it is independent of t, as $\phi_{x,y}$. Of course $\phi_{x,y} = \phi_{x,y}(\theta)$. If $x \equiv y$, $\phi_{x,x}$ is called auto-correlation function. $\phi_{x,y}$ is the Fourier transform of a time-independent bounded measure $\tilde{\omega}_{x,y}$:

$$\phi_{x,y} = \mathscr{F}(\tilde{\omega}_{x,y}) \tag{4.21}$$

and in particular

$$\phi_{x,x} = \mathscr{F}(\tilde{\omega}_{x,x}) \tag{4.22}$$

where $\tilde{\omega}_{x,x}$ is a bounded positive measure. $\tilde{\omega}_{x,x}$ ($\tilde{\omega}_{x,y}$ respectively) is called spectral power distribution (mutual) of x (x and y). If X, Y are not centered, one defines

$$\phi_{x,y} = \phi_{\bar{x},\bar{y}} \quad , \quad \tilde{\omega}_{x,y} = \tilde{\omega}_{\bar{x},\bar{y}} \quad . \tag{4.23}$$

Notice that

$$\int \tilde{\omega}_{x,y} = v(x(t), y(t)) \tag{4.24}$$

and

$$\int \tilde{\omega}_{x,x} = \sigma^2(x(t)) \quad . \tag{4.25}$$

If $\tilde{\omega}_{xx}$ (resp. $\tilde{\omega}_{xy}$) admits a density Θ_{xx} (resp. Θ_{xy}) with respect to the Lebesgue measure, the latter will be referred to as spectral power density (mutual) of x (x and y).

Given the scalar random process X stationary of the second order, and the associated unitary representation $\theta \to U_\theta$, we define — for any bounded measure μ — the process $Z : t \to z(t)$ according to

$$z(t) = \int x(t-\theta) \cdot \mu^\theta \equiv U[\mu]x(t) \quad . \tag{4.26}$$

We write in this case

$$Z = \mu \circ X \quad . \tag{4.27}$$

Both the processes Z and $t \to \begin{pmatrix} x(t) \\ z(t) \end{pmatrix}$ are stationary of the second order. Indeed one has

$$U[\mu]x(t) = \int U_\theta x(t) \cdot \mu^\theta \tag{4.28}$$

whence

$$z(t-\tau) = U[\mu]x(t-\tau) = U[\mu]U_\tau x(t) =$$

$$= U_\tau U[\mu]x(t) = U_\tau z(t) \quad . \tag{4.29}$$

Notice that $z(t) \in \mathscr{H}_x$, $\forall t \in \mathbb{R}$. Definitions (4.26), (4.27) lead to the following interesting properties. Let μ, ν be two bounded measures, and $t \to \begin{pmatrix} x(t) \\ y(t) \end{pmatrix}$ a random process stationary of the second order, then

$$(\mu \circ \nu) \circ X = \mu \circ (\nu \circ X) \tag{4.30}$$

$$\phi_{x,\nu\circ y} = \nu \circ \phi_{xy} \tag{4.31}$$

$$\tilde{\omega}_{x,\nu\circ y} = \mathscr{F}^{-1}(\nu) \cdot \tilde{\omega}_{x,y} \quad . \tag{4.32}$$

Equation (4.30) can be proved noticing that by the definition of measure convolution

$$[(\mu \circ \nu) \circ x](t) = U[\mu \circ \nu]x(t) = U[\mu]U[\nu]x(t) =$$

$$= [\mu \circ (\nu \circ x)](t) \quad . \tag{4.33}$$

On the other hand, by (4.20) and (4.28)

$$\phi_{x,\nu\circ y} = \phi_{x(t),U[\nu]y(t)} = \nu \circ \phi_{x(t),y(t)} = \nu \circ \phi_{x,y} \tag{4.34}$$

which proves (4.31). Finally, by (4.21) and (4.31) and the properties of the Fourier transform of a convolution product,

$$\tilde{\omega}_{x,\nu\circ y} = \mathscr{F}^{-1}(\phi_{x,\nu\circ y}) = \mathscr{F}^{-1}(\nu) \cdot \phi_{x,y} \tag{4.35}$$

which is nothing but (4.32).

A special case of (4.32) is

$$\tilde{\omega}_{\mu\circ x,\mu\circ x} = \left|\mathscr{F}^{-1}(\mu)\right|^2 \cdot \tilde{\omega}_{x,x} \quad . \tag{4.36}$$

Moreover,

$$m_{\mu\circ x} = m_x \int \mu \tag{4.37}$$

Indeed by (4.2), (4.18)

$$m_{\mu\circ x} = E[(\mu \circ x)(t)] = \langle 1, (\mu \circ x)(t) \rangle =$$

$$= \int \langle 1, x(t) \rangle \cdot \mu^t = \int m_x \cdot \mu^t \tag{4.38}$$

which gives (4.37) since m_X is independent on t. Equation (4.37) can be written as well

$$m_{\mu\circ X} = m_X \cdot (\mathscr{F}^{-1}(\mu))(0) \quad . \tag{4.39}$$

Equation (4.37) leads to another generalization. Let X be a random process, stationary of the second order, centered or not. For any measure μ one can define the random variable x_μ by

$$x_\mu = \int x(t)\cdot\mu^t = (\hat{\mu}\circ x)(0) = U[\hat{\mu}]x(0) \tag{4.40}$$

where the hat ^ denotes the involution corresponding to the change of sign of the argument. Computing the mean of such a random variable, one gets

$$E(x_\mu) = \langle 1, x_\mu \rangle = \int \langle 1, x(t) \rangle \cdot \mu^t =$$

$$= \int E(x(t))\cdot\mu^t = m_X \int \mu = m_{\mu\circ X} \quad . \tag{4.41}$$

Moreover

$$v(x_\nu, x_\mu) = \int \tilde{\omega}_{\hat{\nu}\circ X, \hat{\mu}\circ X} =$$

$$= \int [\mathscr{F}^{-1}(\nu)]^* \cdot [\mathscr{F}^{-1}(\mu)] \cdot \tilde{\omega}_{X,X}$$

$$= \int [\mathscr{F}(\nu)]^* \cdot [\mathscr{F}(\mu)] \cdot \tilde{\omega}_{X,X} \quad . \tag{4.42}$$

On the other hand,

$$\langle x_\mu, x_\nu \rangle = \phi_{\hat{\nu}\circ X, \hat{\mu}\circ X}(0) = (\nu^* \circ \hat{\mu} \circ \phi_{X,X})(0) =$$

$$= \int \phi_{XX} \cdot (\bar{\nu}\circ\mu) . \tag{4.43}$$

The bar in (4.43) denotes the involution corresponding to complex conjugation plus the change of sign of the argument.

From (4.42), (4.43) one gets the relevant identity

$$\int [\mathscr{F}(\nu)]^* \cdot \mathscr{F}(\mu) \cdot \tilde{\omega}_{xx} = \int \phi_{xx} \cdot (\bar{\nu} \circ \mu) \quad . \tag{4.44}$$

As particular cases of previous formulas, we have

$$E(x_{\delta_a \circ \mu}) = E(x_\mu) \tag{4.45}$$

$$v(x_{\delta_a \circ \nu}, x_{\delta_a \circ \mu}) = v(x_\nu, x_\mu) \qquad \forall\, a \in \mathbb{R} \quad . \tag{4.46}$$

Moreover we can rewrite all of them when the measures μ, ν read $f(t)dt$, $g(t)dt$, with $f, g \in L^1$. We have

$$v\left(\int g(t)x(t)dt, \int f(t)x(t)dt\right) = \int [\mathscr{F}(g)]^* \mathscr{F}(f) \cdot \tilde{\omega}_{xx} =$$

$$= \int \bar{g} \circ f \cdot \phi_{xx} \tag{4.47}$$

$$E\left(\int f(t)x(t)dt\right) = m_x \int f(t)dt \tag{4.48}$$

instead of (4.44) and (4.41) respectively, and

$$E\left(\int (\delta_a \circ f)(t)x(t)dt\right) = E\left(\int f(t)x(t)dt\right) \tag{4.49}$$

and

$$v\left(\int (\delta_a \circ g)(t)x(t)dt, \int (\delta_a \circ f)(t)x(t)dt\right) =$$

$$= v\left(\int f(t)x(t)dt, \int g(t)x(t)dt\right) \qquad \forall\, a \in \mathbb{R} \tag{4.50}$$

instead of (4.45) and (4.46) respectively.

A typical application of these concepts is the following. Let $\mathscr{D}'_+(\mathscr{V})$ denote the space of distributions on the vector space $\mathscr{V}$ with positive support, and D the derivative of a distribution. Recall that for $T \in \mathscr{D}'_+$, DT is defined by

$$\int f \cdot DT = - \int f' \cdot T \quad , \quad \forall f \in \mathscr{D} \tag{4.51}$$

where $\mathscr{D}$ is the vector space of infinitely derivable complex functions with compact support, and f' denotes the usual derivative of f. Equation (4.51) can be iterated any number of times, to define the p-th derivative D^pT of T:

$$\int f \cdot D^pT = (-)^p \int f^{(p)} \cdot T \tag{4.52}$$

where $f^{(p)}$ is the p-th derivative of f. Let moreover $P(D)$ denote a formal polynomial in D, where the p-th power of D is identified with D^p. Thus

$$P(D)z = x(t) \tag{4.53}$$

is a differential equation. Let $z_0(t)$ denote the elementary solution in $\mathscr{D}'_+$ of (4.53). If $x(t)$ has positive support, the general solution of (4.53) with positive support is

$$z(t) = (z_0 \circ x)(t) = \int_0^t z_0(t-\theta)x(\theta)d\theta \quad , \quad \forall t \geq 0 \quad . \tag{4.54}$$

Let now X be a random process, stationary of the second order, H the Heaviside step function and define $y(t)$ as the response to the equation

$$P(D)y = H(t)\, x(t) \quad . \tag{4.55}$$

We have once more

$$y(t) = \int_0^t z_0(t-\theta)\, x(\theta)\, d\theta \qquad \forall\, t \geq 0 \quad . \tag{4.56}$$

Upon setting, for $t > 0$

$$h_t(\theta) = \begin{cases} z_0(t-\theta) & \text{for } \theta \in [0,t] \\ 0 & \text{for } \theta \notin [0,t] \end{cases} \tag{4.57}$$

one has

$$y(t) = \int x(\theta) h_t(\theta) d\theta = \int h_t \cdot x \tag{4.58}$$

whence, for $t_1, t_2 \geq 0$

$$v(y(t_1), y(t_2)) = \int [\mathscr{F}(h_{t_1})]^* \cdot [\mathscr{F}(h_{t_2})] \cdot \tilde{\omega}_{xx} \quad . \tag{4.59}$$

The process Y is obviously non-stationary.

In terms of the above elements, one can now introduce still a further generalization of the concept of random process. We have seen how any random process, stationary of the second order, associates with every function $f \in L^1$, a random variable $\int f \cdot x \in \mathscr{H}$. It is natural now to restrict the domain of such a correspondence to $\mathscr{D}$. Adding to this the condition of continuity, the mapping $f \to \int f \cdot x$ is defined as a distribution valued in $\mathscr{H}$. We define <u>process-distribution</u> any tempered distribution x on $\mathbb{R}$ valued in the space of square integrable random variables $\mathscr{H}$, namely any continuous map $\nu : f \to \int f \cdot x$ of $\mathscr{S}$ onto $\mathscr{H}$ (where $\mathscr{S}$ is the space of infinitely derivable functions f whose derivatives decrease asymptotically faster than any power: $x^k f^{(h)}(x) \to 0$ for $x \to \infty$, $\forall k, h \in \mathbb{Z}^{(+)}$). All the concepts defined for random processes can be extended in a straightforward way to process-distributions. On the other hand, the convolution by an element of $\mathcal{S}$ transforms any process-distribution into a random process.

Indeed, let $\xi \in \mathcal{H}$ and $f, g, h \in \mathcal{S}$; and consider

$$<\xi, \int f(t)\left[\int h(t-\theta)\cdot x^\theta\right] dt> =$$

$$= \int f(t) <\xi, \int h(t-\theta)\cdot x^\theta> dt \quad . \tag{4.60}$$

Upon introducing the notation

$$<\xi, \int g \cdot x> = \int g \cdot x_\xi \tag{4.61}$$

(4.60) reads

$$\int f(t)\left[\int h(t-\theta)\cdot x_\xi^\theta\right] dt \equiv \int f(t)(h*x_\xi)(t) dt =$$

$$= \int f(t)(h*x)_\xi(t) = <\xi, \int f \cdot h*x> \quad . \tag{4.62}$$

Comparing (4.62) with (4.60) one finally gets

$$\int f(t)\left[\int h(t-\theta)\cdot x^\theta\right] dt = \int f \cdot h*x \tag{4.63}$$

in other words, $h*x$ is a random process, with

$$(h*x)(t) = \int h(t-\theta)\cdot x^\theta \quad . \tag{4.64}$$

To conclude, we consider the application of a special type of random process (of particular interest in physics): the Poisson process. It refers to the phenomena of the same nature (e.g. particle emission or absorption) taking place in time at discrete instants whose distribution is casual. Let us denote by $n(T)$ the number of occurrences of the phenomenon in the time interval T, where T is of the form $[t_i, t_f[$, t_i and t_f being finite. The corresponding process is Poisson if the random variable $n(T)$ satisfies the following conditions:

i) For every interval T the probability distribution $n(T)$ is defined on a suitable space $\mathscr{V}$.

ii) $n(T)$ is a Poisson probability distribution of parameter ℓc, where ℓ is the length of the interval T and c a positive constant — referred to as the process density. In other words, by [(4.25), Chap. 3]

$$\text{prob}\{n(T) = p\} = e^{-c\ell} \frac{(c\ell)^p}{p!} \quad . \tag{4.65}$$

iii) If T_1 and T_2 are two disjoint intervals, then $n(T_1)$ and $n(T_2)$ are independent random variables.

From the definition and [(4.27), (4.29), Chap. 3], there follows that

$$E(n(T)) = c\ell \quad , \quad \sigma^2(n(T)) = c\ell \quad . \tag{4.66}$$

We want to construct a process-distribution out of a Poisson process. Let $\mathfrak{F}$ be the vector space of those functions which cannot take but a finite number of values, each non-vanishing value being taken over a finite union of intervals. More explicitly, two functions f, g belong to $\mathfrak{F}$ if their graph has the form exemplified by Fig. 37, namely, one can find a finite number of intervals T_i such that:

$$f(t) = \begin{cases} \xi_i & \text{for} \quad t \in T_i \\ 0 & \text{for} \quad t \notin \bigcup_i T_i \end{cases} \quad , \quad g(t) = \begin{cases} \eta_i & \text{for} \quad t \in T_i \\ 0 & \text{for} \quad t \notin \bigcup_i T_i \end{cases} \quad . \tag{4.67}$$

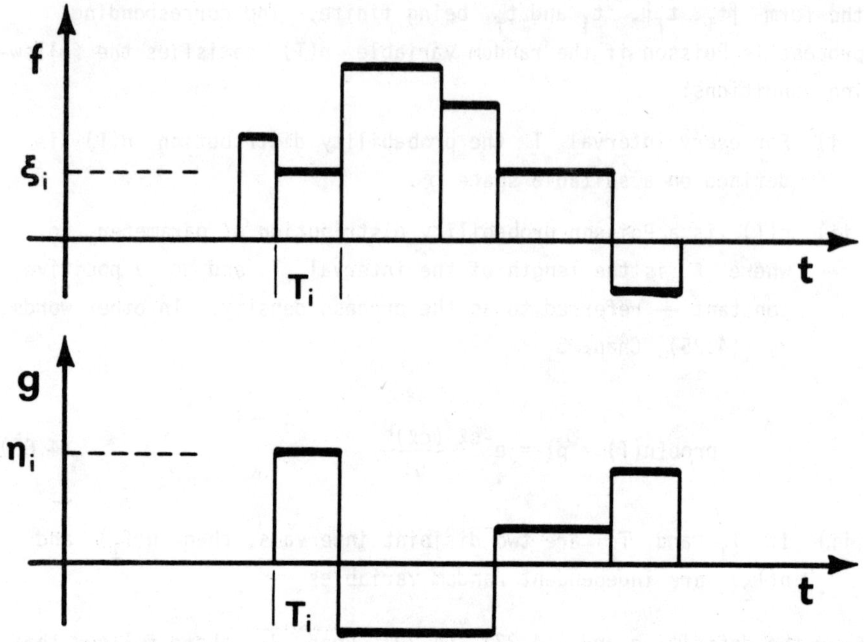

Fig. 37

Let ℓ_i denote the length of the interval T_i. One can associate to the functions $f, g \in \mathfrak{F}$, the random variables

$$x_f = \sum_i \xi_i \; n(T_i) \quad , \quad x_g = \sum_i \eta_i \; n(T_i) \quad , \tag{4.68}$$

one has

$$E(x_f) = \sum_i \xi_i \; E(n(T_i)) = \sum_i \xi_i \; c\ell_i = c \sum_i \xi_i \; \ell_i = c \int f \tag{4.69}$$

and analogously $E(x_g) = c \int g$.

Moreover,

$$v(x_f,x_g) = \sum_{i,j} \eta_i^* \xi_j \, v(n(T_j),n(T_i)) =$$

$$= \sum_j \eta_j^* \xi_j \, \sigma^2(n(T_j)) = \sum_j \eta_j^* \xi_j \, c\ell_j =$$

$$= c \int g^* f \qquad . \tag{4.70}$$

In particular

$$\sigma^2(x_f) = c \int |f|^2 \qquad . \tag{4.71}$$

The norm of x_f in $\mathscr{H}$ is therefore given by:

$$\|x_f\|^2 = c^2 \left| \int f \right|^2 + c \int | f |^2 \qquad . \tag{4.72}$$

One can then consider the Banach space $B = L^1 \cap L^2$, equipped with the norm induced by (4.72):

$$f \to \|f\|_{1,2} = \|f\|_{L^1} + \|f\|_{L^2} \qquad . \tag{4.73}$$

$\mathfrak{F} \subset B$ and the map $f \to x_f$ is a linear continuous map of $\mathfrak{F}$, for $\mathfrak{F}$ equipped with the topology (4.73), into $\mathscr{H}$; which can be extended to a continuous map x of B into $\mathscr{H}$. The latter is in particular defined and continuous over S, and is therefore a process-distribution. Upon defining

$$x_f = \int f \cdot x \tag{4.74}$$

the formulas

$$E\left(\int f \cdot x\right) = c \int f \tag{4.75}$$

$$v\left(\int g \cdot x, \int f \cdot x\right) = c \int g^* f \tag{4.76}$$

valid for $f, g \in \mathfrak{F}$, by continuity hold for $f, g \in B$ as well.

We can now check that the process-distribution thus defined is stationary of the second order. Indeed the requirements that guarantee stationarity read, in the case of a process-distribution:

$$E\left(\int (\delta_a * f) \cdot x\right) = E\left(\int f \cdot x\right) \qquad \forall\, a \in \mathbb{R}$$

and

$$v\left(\int (\delta_a * f) \cdot x, \int (\delta_a * g) \cdot x\right) = v\left(\int f \cdot x, \int g \cdot x\right) \quad . \tag{4.77}$$

In the case under consideration, due to (4.75), (4.76) we have:

$$E\left(\int (\delta_a * f) \cdot x\right) = c\int \delta_a * f = c\int f = E\left(\int f \cdot x\right) \tag{4.78}$$

and

$$v\left(\int (\delta_a * f) \cdot x, \int (\delta_a * g) \cdot x\right) = c\int (\delta_a * f)^* \cdot (\delta_a * g) =$$

$$= c\int g^* f = v\left(\int f \cdot x, \int g \cdot x\right) =$$

$$= \frac{c}{2\pi} \int (\mathscr{F}(g))^* \; (\mathscr{F}(f)) \quad . \tag{4.79}$$

The last step, which simply comes from known properties of the Fourier transform, allows to obtain, by comparison with (4.47), the spectral density of the process as

$$\Theta_{xx} = \frac{c}{2\pi} = \text{const.} \tag{4.80}$$

A Poisson process of density c can be considered as a process-distribution whose spectral measure equals the Lebesgue measure times $c/2\pi$; the correlation distribution is equal to $c \cdot \delta$ and the average $m_x = c$. One usually expresses the fact that a process (such as a Poisson process) has constant spectral density, saying that it is a <u>white noise</u>. From (4.64) it also follows that the convolution $h*x$ of a Poisson process x by a function $h \in B$ is a random process of spectral density $c/2\pi|\mathscr{F}(h)|^2$, auto-correlation function $\tilde{h}*h$ — where $\tilde{h} = \mathscr{F}^{-1}(\mathscr{F}(h)^*)$ — and average $c\int h$.

Chapter 5
A Few Elements of Quantitative Information Theory

1. The Amount of Information

The attempt to construct a quantitative theory in which the concepts of "production" and "transmission" of information are meaningful faces two main difficulties. The definition of a mathematical model in which such concepts are well defined and the assignment of a measure to the amount of information involved. Hereafter the basic elements of such a theory will be reviewed in a way which naturally lends itself to application in physics.

It is assumed that one receives information whenever one "learns" of an event whose occurrence was previously uncertain. Moreover, the more likely an event is, the less information is conveyed by the knowledge of its actual occurrence. Let x represent [the occurrence of] an event, and $\bar{x}$ its complement (i.e. its non-occurrence), and let p_x, $p_{\bar{x}}$, where

$$p_x + p_{\bar{x}} = 1 \tag{1.1}$$

denote the probabilities of two such events. Furthermore, let I_x denote the amount of information conveyed by the knowledge of the occurrence of x. Assuming that x is specified only by its probability p_x, let us define I_x to be a non-negative function I of p_x, defined over the range $0 < p_x \leq 1$,

$$I_x = I(p_x) \tag{1.2}$$

(notice that $p_x = 0$ is meaningless in the present context). Since the probability of receiving the amount of information I_x is p_x — and that of receiving $I_{\bar{x}}$ is $p_{\bar{x}}$ — the expected amount of information received is

$$H(x,\bar{x}) = p_x I(p_x) + p_{\bar{x}} I(p_{\bar{x}}) \quad . \tag{1.3}$$

Of course the symbolism can be easily generalized to more complex cases: if $\{x_1,\ldots,x_n\}$ are a set of mutually exclusive events of probabilities $\{p_{x_1},\ldots,p_{x_n}\}$ respectively, such that

$$p_{x_1} + \ldots + p_{x_n} = 1 \tag{1.4}$$

then the average amount of information conveyed by the knowledge of which x_i actually occurred is assumed to be

$$H(x_1,\ldots,x_n) = p_{x_1} I(p_{x_1}) + \ldots + p_{x_n} I(p_{x_n}) \tag{1.5}$$

(notice that if $p_{x_j} = 0$ for some j, the event x_j should simply be omitted from consideration).

It is interesting that this intuitive type of reasoning leads to strong constraints upon the form of $I(\ )$. For simplicity let us begin with $n=3$ (and set $x_1 = x$, $x_2 = y$, $x_3 = z$). In order to determine which among the events x, y, z actually occurred, it is e.g. sufficient to determine whether or not x occurred, and in the case where it didn't, to determine which of y, z did occur. The amount of information conveyed by the first determination is evidently

$$H(x,\bar{x}) = p_x I(p_x) + (1-p_x) I(1-p_x) \quad . \tag{1.6}$$

If x didn't occur, then the conditional probabilities of y and z are respectively $p_y/p_{\bar{x}}$, $p_z/p_{\bar{x}}$. The amount of information conveyed by the second determination is therefore given by

$$H(y,z|\bar{x}) = \frac{p_y}{p_{\bar{x}}} I\left(\frac{p_y}{p_{\bar{x}}}\right) + \frac{p_z}{p_{\bar{x}}} I\left(\frac{p_z}{p_{\bar{x}}}\right) \quad . \tag{1.7}$$

However this latter amount of information is conveyed only when the event $\bar{x}$ occurs (i.e. x does not occur) so the total amount of information conveyed on the average, by the two determinations is

$$H(x,y,z) = H(x,\bar{x}) + p_{\bar{x}} H(y,z|\bar{x}) \quad . \tag{1.8}$$

The requirement to be imposed on the function $I(\ )$ is that the relation (1.8) be satisfied for all allowable values of p_x, p_y, p_z greater than zero. Explicitly (1.8) writes, expressing the H-functions in terms of probabilities,

$$H(p_x,p_y,p_z) = H(p_x,1-p_x) + (1-p_x)\, H\left(\frac{p_y}{1-p_x}, \frac{p_z}{1-p_x}\right) \quad . \tag{1.9}$$

Now the identical reasoning applies if x is considered to be a composite event, namely if x is assumed to consist of $(n-1)$ mutually exclusive events $u_1,\dots,u_{n-1}$, whose probabilities are denoted, $p_1,\dots,p_{n-1}$. Writing $q_1 \equiv p_y$, $q_2 \equiv p_z$, and $p_n \equiv p_{\bar{x}}$, $(p_n = q_1 + q_2)$, (1.9) reads in this case

$$H(p_1,\dots,p_{n-1},q_1,q_2) = H(p_1,\dots,p_n) + p_n H\left(\frac{q_1}{p_n}, \frac{q_2}{p_n}\right) \quad . \tag{1.10}$$

Condition (1.10) is very strong; indeed it will practically suffice to determine the form of $H(p_1,\dots,p_n)$ without regard to its definition in terms of $I(\ )$. The latter will only impose a further condition suggested by the fact that terms of the form $p_i I(p_i)$ should be dropped when $p_i = 0$, namely that $H(p_1,\dots,p_n)$ be defined even when some of the p_i's vanish but that it be continuous in the domain

$$\mathscr{D}: \{p_i \ge 0 \ ; \ i = 1,\dots,n \mid \sum_{i=1}^{n} p_i = 1\} \quad . \tag{1.11}$$

The function $H(p_1,\ldots,p_n)$ is completely determined up to a multiplicative constant — which fixes the size of the unit of information — by the only additional requirement that it must be a symmetric function of all its variables.

In order to explicitly construct H, let us begin with a few remarks. First, writing (1.9) with $p_x = p_y = 1/2$, $p_z = 0$, one has

$$H\left(\frac{1}{2},\frac{1}{2},0\right) = H\left(\frac{1}{2},\frac{1}{2}\right) + \frac{1}{2}H(1,0) \quad . \tag{1.12}$$

Using now the hypothesis of complete symmetry, one has

$$H\left(\frac{1}{2},\frac{1}{2},0\right) = H\left(0,\frac{1}{2},\frac{1}{2}\right) \tag{1.13}$$

and finally using (1.9) again with $p_x = 0$, $p_y = p_z = 1/2$,

$$H\left(0,\frac{1}{2},\frac{1}{2}\right) = H(0,1) + H\left(\frac{1}{2},\frac{1}{2}\right) \quad . \tag{1.14}$$

Inserting (1.12) and (1.14) into (1.13) and recalling that one should have $H(0,1) = H(1,0)$, (1.13) implies

$$H(1,0) = 0 \quad . \tag{1.15}$$

Equation (1.15) in turn implies that

$$H(p_1,\ldots,p_n,0) = H(p_1,\ldots,p_n) \tag{1.16}$$

as one can straightforwardly derive from (1.10), setting in it $q_2 = 0$ (and hence $q_1 = p_n$). One can now by induction extend (1.10) to ($m \geq 2$)

$$H(p_1,\ldots,p_{n-1},\, q_1,\ldots,q_m) = H(p_1,\ldots,p_n) + p_n H\left(\frac{q_1}{p_n},\ \ldots\ ,\frac{q_m}{p_n}\right) \tag{1.17}$$

where

$$p_n = q_1 + \ldots + q_m \quad . \tag{1.18}$$

Notice that (1.17) indeed coincides with (1.10) for $m = 2$. From (1.16), it is clear that one has to consider only the case when $q_i > 0$, $i = 1,\ldots,m$. Suppose there is an m such that (1.17), (1.18) are true for all n, then, setting

$$p' = q_2 + \ldots + q_{m+1} \tag{1.19}$$

and using (1.10) as well one can write

$$\begin{aligned} H(p_1,\ldots,p_{n-1}, q_1,\ldots,q_{m+1}) &= \\ &= H(p_1,\ldots,p_{n-1}, q_1,p') + p'H\left(\frac{q_2}{p'},\ldots,\frac{q_{m+1}}{p'}\right) = \\ &= H(p_1,\ldots,p_n) + p_nH\left(\frac{q_1}{p_n},\frac{p'}{p_n}\right) + p'H\left(\frac{q_2}{p'},\ldots,\frac{q_{m+1}}{p'}\right) \quad . \end{aligned} \tag{1.20}$$

But, once more for the induction hypothesis (to be thought of for $n = 2$),

$$H\left(\frac{q_1}{p_n},\ldots,\frac{q_{m+1}}{p_n}\right) = H\left(\frac{q_1}{p_n},\frac{p'}{p_n}\right) + \frac{p'}{p_n}H\left(\frac{q_2}{p'},\ldots,\frac{q_{m+1}}{p'}\right) \tag{1.21}$$

i.e.

$$p_nH\left(\frac{q_1}{p_n},\frac{p'}{p_n}\right) + p'H\left(\frac{q_2}{p'},\ldots,\frac{q_{m+1}}{p'}\right) = p_nH\left(\frac{q_1}{p_n},\ldots,\frac{q_{m+1}}{p_n}\right) \tag{1.22}$$

which, inserted at the right-hand side of (1.20) gives the assertion of (1.17) for $m + 1$.

It is now easy matter to prove that if

$$p_i = \sum_{j=1}^{m_i} q_{i,j} \quad , \qquad i = 1,\dots,n \tag{1.23}$$

then

$$H(q_{1,1},\dots,q_{1,m_1},\dots,q_{n,1},\dots,q_{n,m_n}) =$$

$$= H(p_1,\dots,p_n) + \sum_{i=1}^{n} p_i H\left(\frac{q_{i,1}}{p_i},\dots,\frac{q_{i,m_i}}{p_i}\right) . \tag{1.24}$$

In fact, using (1.17), one has

$$H(q_{1,1},\dots,q_{n,m_n}) = p_n H\left(\frac{q_{n,1}}{p_n},\dots,\frac{q_{n,m_n}}{p_n}\right) =$$

$$+ H(q_{1,1},\dots,q_{n-1,m-1},p_n) \qquad . \tag{1.25}$$

After shifting p_n in the last factor to the extreme left by the assumed symmetry of H, the reduction process can be repeated on it. After n steps the result (1.24) is obtained. Let us now set

$$F(n) \doteqdot H\left(\frac{1}{n},\dots,\frac{1}{n}\right) \quad , \qquad n \geq 2 \tag{1.26}$$

and $F(1) = 0$. Applying (1.24) to the case $m_1 = \dots = m_n = m$, $q_{ij} = 1/mn$, $\forall\, i,j = 1,\dots,n$, one has

$$F(mn) = F(m) + F(n) \qquad . \tag{1.27}$$

Applying further (1.17) one obtains

$$H\left(\frac{1}{n},\dots,\frac{1}{n}\right) = H\left(\frac{1}{n},\frac{n-1}{n}\right) + \frac{n-1}{n} H\left(\frac{1}{n-1},\dots,\frac{1}{n-1}\right) \tag{1.28}$$

from which

$$\eta_n \doteq H\left(\frac{1}{n}, \frac{n-1}{n}\right) = F(n) - \frac{n-1}{n} F(n-1) \quad . \tag{1.29}$$

Now from the continuity of $H(p,1-p)$ there follows that as $n \to \infty$, $\eta_n \to H(0,1) = 0$ (see (1.15)). Moreover the recursion relation

$$n\,\eta_n = nF(n) - (n-1)F(n-1) \tag{1.30}$$

implies

$$nF(n) = \sum_{k=1}^{n} k \cdot \eta_k \tag{1.31}$$

or

$$\frac{F(n)}{n} = \frac{1}{n^2} \sum_{k=1}^{n} k \cdot \eta_k = \frac{n+1}{2n} \frac{2}{n(n+1)} \sum_{k=1}^{n} k \cdot \eta_k \quad . \tag{1.32}$$

But $\frac{2}{n(n+1)} \sum_{k=1}^{n} k \cdot \eta_k$ is simply the arithmetic mean of the first $\frac{n(n+1)}{2}$ terms of the sequence $\eta_1, \eta_2, \eta_2, \eta_3, \eta_3, \eta_3, \eta_4, \eta_4, \eta_4, \eta_4, \ldots$ whose limit, as it was shown above is zero. Thus as $n \to \infty$, $\frac{2}{n(n+1)} \sum_{k=1}^{n} k \cdot \eta_k \to 0$, from which follows that also $\frac{F(n)}{n} \to 0$.

Finally let

$$\lambda_n \equiv F(n) - F(n-1) = \eta_n - \frac{1}{n} F(n-1) \tag{1.33}$$

$\lambda_n \to 0$ as $n \to \infty$, as well. We have thus all the ingredients needed to determine the form of $F(n)$. It is clear from (1.27) that we only need to know the value of $F(n)$ for n prime. Indeed, for arbitrary n, let

$$n = p_1^{\alpha_1} \ldots p_s^{\alpha_s} \quad , \quad s \geq 1 \quad , \quad \alpha_s \in \mathbb{Z}_+ \tag{1.34}$$

be the prime factorization of n. Repeated application of (1.27) gives then

$$F(n) = \alpha_1 F(p_1) + \dots + \alpha_s F(p_s) \quad . \tag{1.35}$$

We now put, for all prime p

$$F(p) = c_p \ell n\, p \tag{1.36}$$

so that (1.35) reads

$$F(n) = \alpha_1 c_{p_1} \ell n\, p_1 + \dots + \alpha_s c_{p_s} \ell n\, p_s \quad . \tag{1.37}$$

It is straightforward to show that the sequence $\{c_p, p = \text{prime}\}$ contains a largest member. In fact, if the contrary were true, it would be possible to construct an infinite sequence of primes $p_1 < p_2 < p_3 < \dots$ with $p_1 = 2$ such that p_{i+1} were the first prime greater than p_i for which $c_{p_{i+1}} > c_{p_i}$. There would follow from this construction that if q is a prime less than p_i, then $c_q < c_{p_i}$. Now, for $i > 1$, let $p_i - 1 = q_1^{\beta_1} \dots q_r^{\beta_r}$, $r \geq 1$, $\beta_r \in \mathbb{Z}_+$ be the prime factorization of $(p_i - 1)$, and consider

$$\begin{aligned}\lambda_{p_i} &= F(p_i) - F(p_i - 1) = \\ &= F(p_i) - \frac{F(p_i)}{\ell n p_i} \ell n(p_i - 1) + c_{p_i} \ell n(p_i - 1) - F(p_i - 1) = \\ &= \frac{F(p_i)}{p_i} \frac{p_i}{\ell n p_i} \ell n \frac{p_i}{p_i - 1} + \sum_{j=1}^{r} \beta_j (c_{p_i} - c_{q_j}) \ell n q_j \quad .\end{aligned} \tag{1.38}$$

Since $(p_i - 1)$ is necessarily even, one of the q_j must take on the value 2. Moreover since $p_i > q_j$, $j = 1,\dots,r$, by the hypothesis we should have

$$\sum_{j=1}^{r} \beta_j(c_{p_i} - c_{q_j})\ln q_j \geq (c_{p_i} - c_2)\ln 2 \geq (c_{p_2} - c_2)\ln 2 \quad . \tag{1.39}$$

However, as $i \to \infty$, by the properties proved above, both λ_{p_i} and $\frac{F(p_i)}{p_i}$ tend to zero $\left(\text{and so does } \frac{p_i}{\ln p_i}\ln\frac{p_i}{p_i - 1}\right)$. Therefore, by (1.38), (1.39) it should be $(c_{p_2} - c_2)\ln 2 \leq 0$ or $c_{p_2} \leq c_2$, which contradicts the definition of p_2. In precisely the same manner one shows that the sequence $\{c_p, p = \text{prime}\}$ contains a smallest member. Suppose now that there is at least a prime $\bar{p}$ such that $c_{\bar{p}} > c_2$, and let p_0 be that prime for which c_{p_0} is a maximum. Of course then $c_{p_0} > c_2$. Let m be a positive integer ≥ 1 and let $\bar{q}_1^{\gamma_1} \ldots \bar{q}_t^{\gamma_t}$ be the prime factorization of $p_0^m - 1$. From (1.27) it follows that

$$\frac{F(p_0^m)}{\ln p_0^m} = c_{p_0} \quad . \tag{1.40}$$

Then we can repeat all the steps leading to (1.38) and (1.39) (β_j being replaced by γ_j, r by t, q_j by $\bar{q}_j$ and p_i by p_0^m) obtaining

$$\lambda_{p_0^m} \geq \frac{F(p_0^m)}{p_0^m}\frac{p_0^m}{\ln p_0^m}\ln\frac{p_0^m}{p_0^m - 1} + (c_{p_0} - c_2)\ln 2 \quad . \tag{1.41}$$

Letting $m \to \infty$, one gets $(c_{p_0} - c_2)\ln 2 \leq 0$, which contradicts $c_{p_0} > c_2$. In precisely the same manner, one shows the non-existence of any prime q_0 for which $c_{q_0} < c_2$. Thus, all the c_p's are equal and (from (1.34), (1.37))

$$F(n) = c \ln n \tag{1.42}$$

where c is a constant (equal) to the common value of the c_p's).

Let finally

$$p = \frac{r}{s} \quad , \quad r,s \in \mathbb{Z}_+ \quad , \quad s \geq r \quad . \tag{1.43}$$

By (1.24) one can write

$$H\left(\frac{1}{s}, \ldots, \frac{1}{s}\right) = H\left(\frac{r}{s}, \frac{s-r}{s}\right) + \frac{r}{s} H\left(\frac{1}{r}, \ldots, \frac{1}{r}\right) + + \frac{s-r}{s} H\left(\frac{1}{s-r}, \ldots, \frac{1}{s-r}\right) \tag{1.44}$$

from which

$$\begin{aligned} H(p\,, 1-p) &= F(s) - pF(r) - (1-p)F(s-r) = \\ &= c\ \ell n\ s - pc\ \ell n\ r - (1-p)c\ \ell n\ (s-r) = \\ &= c\left(p\ \ell n\ \frac{s}{r} + (1-p)\ \ell n\ \frac{s}{s-r}\right) = \\ &= c\left(p\ \ell n\ \frac{1}{p} + (1-p)\ \ell n\ \frac{1}{(1-p)}\right) \quad . \end{aligned} \tag{1.45}$$

By continuity this extends to all irrational p's, and using (1.10) inductively on n, we have

$$H(p_1,\ldots,p_n) = -\ c \sum_{i=1}^{n} p_i\ \ell n\ p_i \quad . \tag{1.46}$$

Notice that $H(p_1,\ldots,p_n)$ in (1.46) is just of the form (1.5),

$$H(p_1,\ldots,p_n) = \sum_{i=1}^{n} p_i\ I(p_i) \quad . \tag{1.47}$$

This suggests that c must be taken > 0, so that

$$I(p) = -\ c\ \ell n\ p \tag{1.48}$$

is a monotone increasing function of $\bar{p} = 1 - p$.

2. Basic Properties of the H Function

Let X be an abstract set, consisting of a finite number of elements x. Let $p(\ )$ be a probability function defined over X, i.e. $p(Q)$ is a non-negative number defined for each subset Q of X, with the properties that:

$$p(X) = 1 \tag{2.1}$$

and

$$p(Q_1 \cup Q_2) = p(Q_1) + p(Q_2) \tag{2.2}$$

if Q_1 and Q_2 are disjoint subsets of X. The totality of objects $\{(X,x),p(\ )\}$ is what was defined as a finite probability space. Recalling the discussion of previous section, any finite probability space can be considered an information source. By obvious extension of (1.46), we will define as information content of such a source the non-negative quantity

$$H(X) = - \sum_{x \in X} p(x)\ \ell n\ p(x) \tag{2.3}$$

(where units chosen in such a way that $c = 1$).

Let now (X,x) and (Y,y) be two finite abstract spaces, and denote by $X \otimes Y$ the finite abstract space consisting of all pairs (x,y), and by $p(\ ,\ )$ a probability distribution over $X \otimes Y$. The information content of this source is obviously

$$H(X \otimes Y) = - \sum_{x \in X} \sum_{y \in Y} p(x,y)\ \ell n\ p(x,y) \quad . \tag{2.4}$$

However in this case the distribution $p(\ ,\ )$ gives rise also to a distribution

$$p(x) = \sum_{y \in Y} p(x,y) \quad \text{over} \quad (X,x) \tag{2.5}$$

and to a distribution

$$p(y) = \sum_{x \in X} p(x,y) \quad \text{over} \quad (Y,y) \quad . \tag{2.6}$$

For both of these, assumed as information sources, the information contents $H(X)$, $H(Y)$ respectively can be defined, according to (2.3), without regard to the origin of $p(x)$ or $p(y)$. Further, for each $y \in Y$ satisfying $p(y) > 0$, the probability

$$p(x|y) = \frac{p(x,y)}{p(y)} \tag{2.7}$$

is a probability distribution over X (and analogously $p(y|x) = \frac{p(x,y)}{p(x)}$ is a probability distribution over Y). We may therefore define a conditional information content

$$H(X|y) = - \sum_{x \in X} p(x|y) \ln p(x|y) \tag{2.8}$$

as well as an average conditional information content

$$H(X|Y) = \sum_{y \in Y} p(y) H(X|y) =$$

$$= - \sum_{y \in Y} \sum_{x \in X} p(x,y) \ln p(x|y) \quad . \tag{2.9}$$

(Of course an $H(x|Y)$ can be defined in analogy with (2.8)).

Clearly all the above concepts can be generalized to an n-tuple of finite abstract spaces for any $n \geq 2$, by noticing that if $\{X_i, i = 1,\ldots,n\}$ is a set of finite abstract spaces, one can define

$$X^{(n)} = \otimes \sum_{i=1}^{n} X_i \tag{2.10}$$

as the set of all vectors $(x^1,\dots,x^n)$ where the i-th component $x^i \in X_i$. X has the property that

$$X^{(n)} = \left\{\bigotimes_{i=1}^{j} X_i\right\} \otimes \left\{\bigotimes_{k=j+1}^{n} X_k\right\} \equiv X^{(j)} \otimes X^{(n-j)} \tag{2.11}$$

for any $1 \le j \le n$. Thus if a probability distribution $p(\ ,\dots,\ )$ is given over $X^{(n)}$, expressions such as $H(X^{(j)}|X^{(n-j)})$ may be well defined since they are of the form $H(X|Y)$ above.

The information content functions satisfy a number of inequalities of interest for the applications. In order to derive them, two easy lemmas are needed first. Let $\{p_i\}$, $\{q_i\}$, $i = 1,\dots,n$ be two sets of non-negative numbers such that

$$\sum_{i=1}^{n} p_i = \sum_{i=1}^{n} q_i = 1 \quad . \tag{2.12}$$

i) We have

$$-\sum_{i=1}^{n} q_i \ln q_i \le -\sum_{i=1}^{n} q_i \ln p_i \tag{2.13}$$

with the equality holding if and only if $p_i = q_i$, $\forall\, i$.

ii) Let $\{a_{ij}\}$, $i,j = 1,\dots,n$ be a set of non-negative numbers, such that

$$\sum_{i=1}^{n} a_{ij} = \sum_{j=1}^{n} a_{ij} = 1 \tag{2.14}$$

and let

$$p_i' = \sum_{j=1}^{n} a_{ij}\, p_j \tag{2.15}$$

then we have

$$-\sum_{i=1}^{n} p_i' \,\ell n\, p_i' \geq -\sum_{i=1}^{n} p_i \,\ell n\, p_i \tag{2.16}$$

with the equality holding if and only if the $\{p_i'\}$ are simply a relabelling of the $\{p_i\}$.

In order to prove i), one simply has to resort to the so-called "inequality of the arithmetic and geometric means" (see the appendix at the end of present section for an extended statement of such fundamental inequality), which in present case reads [set in (A.15), $a_i \doteqdot p_i/q_i$, $p_i \Leftrightarrow q_i$, $i = 1,\ldots,n$; and hence by (2.12) $P_n = 1$]

$$\left(\frac{p_1}{q_1}\right)^{q_1}\cdots\left(\frac{p_n}{q_n}\right)^{q_n} \leq 1 \tag{2.17}$$

with the equality holding if and only if $p_i = q_i$, $i = 1,\ldots,n$. Taking the logarithm of both sides of (2.17) one has

$$\sum_{i=1}^{n} q_i \,\ell n\left(\frac{p_i}{q_i}\right) \leq 0 \tag{2.18}$$

from which (2.13) follows at once.

In order to prove ii), one notices first that, quite trivially, one has

$$\begin{aligned} -\sum_{i=1}^{n} p_i' \,\ell n\, p_i' &= -\sum_{i=1}^{n}\sum_{j=1}^{n} a_{ij}\, p_j \,\ell n\, p_i' = \\ &= -\sum_{j=1}^{n} p_j \sum_{i=1}^{n} a_{ij} \,\ell n\, p_i' = \\ &= -\sum_{j=1}^{n} p_j \,\ell n \prod_{i=1}^{n} (p_i')^{a_{ij}} \quad . \end{aligned} \tag{2.19}$$

Now, once more from the inequality of the arithmetic and geometric means, which we now write [set in (A.17), $\frac{A_{ik}}{\sum_{k=1}^{m} A_{ik}} \doteq p_i'$, any k; $1 \leq k \leq m$; and $\lambda_i = a_{ij}$, for each j; $1 \leq j \leq n$; $\forall i$] as a set of n inequalities

$$\prod_{i=1}^{n} (p_i')^{a_{ij}} \leq \sum_{i=1}^{n} a_{ij} p_i' \quad ; \quad j = 1,\ldots,n \quad , \tag{2.20}$$

taking logarithms of both sides, the last factor in (2.19) satisfies

$$- \sum_{j=1}^{n} p_j \ln \prod_{i=1}^{n} (p_i')^{a_{ij}} \geq - \sum_{j=1}^{n} p_j \ln \left\{ \sum_{i=1}^{n} a_{ij} p_i' \right\} \quad . \tag{2.21}$$

Since, by definitions (2.14), (2.15) and (2.12)

$$\sum_{j=1}^{n} \left(\sum_{i=1}^{n} a_{ij} p_i' \right) = 1 \tag{2.22}$$

we can now apply lemma i) and further extend the inequality (2.21):

$$- \sum_{j=1}^{n} p_j \ln \left\{ \sum_{i=1}^{n} a_{ij} p_i' \right\} \geq - \sum_{j=1}^{n} p_j \ln p_j \quad . \tag{2.23}$$

The sequence (2.19), (2.21) and (2.23) proves lemma ii). Notice that if one is to have equality throughout, one must have

$$\prod_{i=1}^{n} (p_i')^{a_{ij}} = \sum_{i=1}^{n} a_{ij} p_i' = p_j \quad , \quad j = 1,\ldots,n \quad . \tag{2.24}$$

Now the condition for the last equality in (2.24) to hold is that those p_i' whose exponent a_{ij} is not zero must be equal for each fixed j. Indeed let $p(j)$ be their common value. Then, due to (2.14),

$$p_j = \sum_{i=1}^{n} a_{ij}\, p_i' = p(j) \qquad , \qquad j = 1,\dots,n \qquad . \tag{2.25}$$

Thus every one of the p_j is equal to some of the p_i''s. On the other hand, since for each i there must be at least one j for which $a_{ij} > 0$, every one of the p_i''s is equal to some p_j.

Let q be a value taken on by some p_i', and let $\mathscr{A}$ be the set of i's for which $p_i' = q$ (obviously $\mathscr{A}$ is a subset of $\{1,\dots,n\}$). Let moreover $\mathscr{B}$ denote the set of j's (also $\mathscr{B} \subset \{1,\dots,n\}$) such that for each $j \in \mathscr{B}$, $a_{ij} > 0$ for at least one value of $i \in \mathscr{A}$. In other words $a_{ij} = 0$ for $j \in \mathscr{B}$ and $i \notin \mathscr{A}$. Then, for $j \in \mathscr{B}$, (2.14) writes

$$\sum_{i \in \mathscr{A}} a_{ij} = 1 \tag{2.26}$$

whence

$$\sum_{j \in \mathscr{B}} \sum_{i \in \mathscr{A}} a_{ij} = |\mathscr{B}| \tag{2.27}$$

is the number of elements of $\mathscr{B}$. On the other hand $a_{ij} = 0$ also for $i \in \mathscr{A}$, $j \notin \mathscr{B}$; thus for $i \in \mathscr{A}$

$$\sum_{j \in \mathscr{B}} a_{ij} = 1 \tag{2.28}$$

and

$$\sum_{i \in \mathscr{A}} \sum_{j \in \mathscr{B}} a_{ij} = |\mathscr{A}| \qquad . \tag{2.29}$$

Comparing (2.27) and (2.29) one finds that

$$|\mathscr{A}| = |\mathscr{B}| \qquad . \tag{2.30}$$

But if $j \in \mathscr{B}$, by (2.25),

$$p_j = \sum_{i \in \mathscr{A}} a_{ij}\, p_i' = q \tag{2.31}$$

thus the $\{p_i'\}$ are just a relabeling of the $\{p_j\}$.

Having proved the two lemmas i) and ii), the properties of H we want to recall are easily derived. Take in ii), $a_{ij} = 1/n$ for all $i,j = 1,\dots,n$. Then, by (2.15) and (2.12), $p_1' = \dots = p_n' = 1/n$. The lemma (e.g. (2.16)) implies then that the maximum information content of a source having n elements is $\ell n\, n$, and is achieved only when all elements have equal probability $p_i = 1/n$, $i = 1,\dots,n$.

Consider now

$$H(X) + H(Y) = -\sum_{x \in X}\sum_{y \in Y} p(x,y)[\ell n\, p(x) + \ell n\, p(y)] =$$

$$= -\sum_{x \in X}\sum_{y \in Y} p(x,y)\, \ell n[p(x)p(y)] \quad . \tag{2.32}$$

Since by definition (2.1)

$$\sum_{x \in X}\sum_{y \in Y} p(x)p(y) = 1 \tag{2.33}$$

we can apply lemma i), which writes

$$-\sum_{x \in X}\sum_{y \in Y} p(x,y)\, \ell n[p(x)p(y)] \geq -\sum_{x \in X}\sum_{y \in Y} p(x,y)\, \ell n\, p(x,y) \quad . \tag{2.34}$$

Recalling definition (2.4), (2.34) together with (2.32) gives

$$H(X \otimes Y) \leq H(X) + H(Y) \tag{2.35}$$

with the equality holding if and only if

$$p(x,y) = p(x)p(y) \quad . \tag{2.36}$$

An immediate consequence of (2.35), coming from definitions (2.3) to (2.9), which imply that

$$H(X \otimes Y) = H(Y) + H(X|Y) \tag{2.37}$$

and hence

$$H(X) + H(Y) - H(X \otimes Y) = H(X) - H(X|Y) \tag{2.38}$$

is that

$$H(X|Y) \leq H(X) \tag{2.39}$$

with the equality holding if and only if X and Y are probabilistically independent (e.g. (2.36)). Further, (2.35) can be iterated n times to give

$$H(X^{(n)}) \leq H(X_1) + \ldots + H(X_n) \quad . \tag{2.40}$$

One last interesting property is the following. Let

$$X = X_1 \quad , \quad Y = X_2 \otimes \ldots \otimes X_{n-1} \quad , \quad Z = X_n \quad . \tag{2.41}$$

Then

$$H(X|Y \otimes Z) \leq H(X|Y) \tag{2.42}$$

with equality if and only if

$$p(x,z|y) = p(x|y)p(z|y) \quad . \tag{2.43}$$

The proof of (2.42) parallels that of (2.39). Indeed,

$$H(X|Y) - H(X|Y \otimes Z) = - \sum_{x \in X} \sum_{y \in Y} p(x,y) \ln p(x|y) +$$

$$+ \sum_{x \in X} \sum_{y \in Y} \sum_{z \in Z} p(x,y,z) \ln p(x|y,z) =$$

$$= - \sum_{y \in Y} p(y) \left[\sum_{x \in X} p(x|y) \ln p(x|y) \right] +$$

$$+ \sum_{y \in Y} p(y) \left[\sum_{x \in X} \sum_{z \in Z} p(x,z|y) \ln p(x|y,z) \right] \quad . \qquad (2.44)$$

Now, by definition

$$p(x|y) = \sum_{z \in Z} p(x,z|y) \qquad (2.45)$$

and

$$p(x|y,z) = \frac{p(x,z|y)}{p(z|y)} \qquad (2.46)$$

so that (2.44) can be written

$$H(X|Y) - H(X|Y \otimes Z) = \sum_{y \in Y} p(y) \left\{ - \sum_{x \in X} p(x|y) \ln p(x|y) + \right.$$

$$+ \sum_{x \in X} \sum_{z \in Z} p(x,z|y) \ln p(x,z|y) -$$

$$\left. - \sum_{z \in Z} p(z|y) \ln p(z|y) \right\} \quad . \qquad (2.47)$$

The factors in curly brackets are but (see (2.8))

$$H(X|y) + H(Z|y) - H(X \otimes Z|y) \qquad (2.48)$$

for which $\mathscr{y}$ can be considered simply as an irrelevant label attached to all probabilities. One can then apply, for each term corresponding to a given $\mathscr{y} \in Y$ in the sum at the right-hand side of (2.47) the inequality (2.35), from which (2.42) follows in a direct manner when the sum itself is performed.

Appendix
The Inequality of the Arithmetic and Geometric Means

Given the sequences of positive numbers $a = (a_1,\ldots,a_n)$ and $b = (b_1,\ldots,b_n)$, we have the following fundamental inequality (referred to as Cauchy-Schwartz inequality):

$$\left(\sum_{k=1}^{n} a_k b_k\right)^2 \leq \left(\sum_{k=1}^{n} a_k^2\right)\left(\sum_{k=1}^{n} b_k^2\right) \quad . \tag{A.1}$$

The proof of (A.1) is based on the elementary inequality, holding for any $x,y \in \mathbb{R}$,

$$xy \leq \frac{1}{2}(x^2 + y^2) \quad , \tag{A.2}$$

that one writes as

$$|a_k b_k| = \lambda |a_k| \cdot \frac{1}{\lambda} |b_k| \leq \frac{1}{2}\left(\lambda^2 a_k^2 + \frac{1}{\lambda^2} b_k^2\right) \quad . \tag{A.3}$$

Choosing

$$\lambda^2 = \left(\frac{\sum_{k=1}^{n} b_k^2}{\sum_{k=1}^{n} a_k^2}\right)^{1/2} \tag{A.4}$$

and summing both sides of (A.3) over k, from 1 to n, one gets

$$\sum_{k=1}^{n} |a_k b_k| \le \left[\left(\sum_{k=1}^{n} a_k^2\right)\left(\sum_{k=1}^{n} b_k^2\right)\right]^{1/2} \quad . \tag{A.5}$$

Recalling finally that

$$\left|\sum_{k=1}^{n} a_k b_k\right| \le \sum_{k=1}^{n} |a_k b_k| \tag{A.6}$$

leads directly to (A.1). In (A.1) the equality holds if and only if the two sequences are proportional [i.e. there exists a constant c such that $a_i = cb_i$, $\forall\, i = 1,\dots,n$].

For a sequence of positive numbers $a = (a_1,\dots,a_n)$ and one of positive weights $p = (p_1,\dots,p_n)$, and for any real number r, one defines as "weighted mean of order r of a", the quantity

$$M_n^{[r]}(a;p) = \left(\frac{\sum_{i=1}^{n} p_i\, a_i^r}{P_n}\right)^{1/r} \tag{A.7}$$

where

$$P_n = \sum_{i=1}^{n} p_i \quad . \tag{A.8}$$

We prove first the following auxiliary theorem: If $a_1 = \dots = a_n = a_0$, then

$$M_n^{[r]}(a;p) = a_0 \tag{A.9}$$

otherwise $M_n^{[r]}(a;p)$ is a strictly monotone increasing function of r, i.e. for $s < t$

$$M_n^{[s]}(a;p) < M_n^{[t]}(a;p) \quad . \tag{A.10}$$

In order to prove (A.10), let us consider $M_n^{[t]}(a;p)$ as a function of t alone, say $f(t)$. By definition $f(t)$ is a strictly positive function; $f(t) > 0$.

We want to prove that $f'(t) = df/dt$ is strictly positive as well. Define

$$F(t) = t^2 \frac{f'(t)}{f(t)} = t \frac{\sum_{i=1}^{n} p_i a_i^t \ln a_i}{\sum_{i=1}^{n} p_i a_i^t} - \ln\left(\frac{\sum_{i=1}^{n} p_i a_i^t}{P_n}\right) . \quad (A.11)$$

Obviously $F(t)$ and $f'(t)$ have the same sign ($\frac{t^2}{f(t)}$ being positive), and we should then prove that $F(t) > 0$. Now, from (A.11),

$$F'(t) = \frac{t}{\left(\sum_{i=1}^{n} p_i a_i^t\right)^2} \left\{\left(\sum_{i=1}^{n} p_i a_i^t\right)\left(\sum_{i=1}^{n} p_i a_i^t \ln^2 a_i\right) - \left(\sum_{i=1}^{n} p_i a_i^t \ln a_i\right)^2\right\} . \quad (A.12)$$

By the Cauchy-Schwartz inequality (A.1), the term in curly brackets on the right-hand side of (A.12) is positive, so $F'(t)$ has the same sign as t. In other words $F(t)$ is increasing for $t > 0$, decreasing for $t < 0$ and hence it has a minimum for $t = 0$. Thus $F(t)$ is positive for all t except possibly $t = 0$. On the other hand, for $t = 0$

$$f'(0) = \left(\prod_{i=1}^{n} a_i^{p_i}\right)^{1/P_n} \cdot$$

$$\cdot \frac{\left(\sum_{i=1}^{n} p_i\right)\left(\sum_{i=1}^{n} p_i \ln^2 a_i\right) - \left(\sum_{i=1}^{n} p_i \ln a_i\right)^2}{2 P_n^2} . \quad (A.13)$$

The numerator at the right-hand side of (A.13) is positive, once more due to Cauchy-Schwartz inequality except when $a_i = a_0$, $\forall\, i = 1,\dots,n$, where $f'(0) = 0$. This completes the proof of (A.10).

Now the Inequality of the Arithmetic and Geometric Means is stated in the theorem: If $\{\lambda_1,\dots,\lambda_n\}$ and $\{A_{ji};\ j = 1,\dots,n;\ i = 1,\dots,m\}$ are sequences of strictly positive numbers $(\lambda_i > 0,\ A_{ji} > 0,\ \forall\, i,j)$ such that $\lambda_1 + \dots + \lambda_n \geq 1$, then

$$\sum_{i=1}^{m} A_{1i}^{\lambda_1} \cdots A_{ni}^{\lambda_n} \leq \left(\sum_{i=1}^{m} A_{1i}\right)^{\lambda_1} \cdots \left(\sum_{i=1}^{m} A_{ni}\right)^{\lambda_n} \quad . \tag{A.14}$$

Consider the case $\lambda_1 + \dots + \lambda_n = 1$ first. To prove (A.14) in this case, we write (A.10) with $s = 0$, $t = 1$:

$$M_n^{[0]}(a;p) = \left(\prod_{i=1}^{n} a_i^{p_i}\right)^{1/P_n} \leq \frac{1}{P_n} \sum_{i=1}^{n} p_i\, a_i = M_n^{[1]}(a;p) \tag{A.15}$$

and replace in it $p_j/P_n = \lambda_j$, $j = 1,\dots,n$ and set for any $1 \leq \nu \leq m$,

$$a_j = \frac{A_{j\nu}}{\sum_{i=1}^{m} A_{ji}} \quad , \quad j = 1,\dots,n \quad . \tag{A.16}$$

We get thus

$$\left(\frac{A_{1\nu}}{\sum_{i=1}^{m} A_{1i}}\right)^{\lambda_1} \cdots \left(\frac{A_{n\nu}}{\sum_{i=1}^{m} A_{ni}}\right)^{\lambda_n} \leq \lambda_1 \frac{A_{1\nu}}{\sum_{i=1}^{m} A_{1i}} + \dots + \lambda_n \frac{A_{n\nu}}{\sum_{i=1}^{m} A_{ni}} \quad . \tag{A.17}$$

Summing now over ν, from 1 to m, one obtains

$$\frac{\sum_{i=1}^{m} A_{1i}^{\lambda_1} \ldots A_{ni}^{\lambda_n}}{\left(\sum_{i=1}^{m} A_{1i}\right)^{\lambda_1} \ldots \left(\sum_{i=1}^{m} A_{ni}\right)^{\lambda_n}} \leq \lambda_1 + \ldots + \lambda_n = 1 \tag{A.18}$$

which is equivalent to (A.14). If now

$$\lambda_1 + \ldots + \lambda_n = k > 1 \tag{A.19}$$

define

$$\mu_j \doteqdot \frac{1}{k} \lambda_j \quad , \quad j = 1,\ldots,n \tag{A.20}$$

so that

$$\mu_1 + \ldots + \mu_n = 1 \tag{A.21}$$

and

$$B_{ji} \doteqdot A_{ji}^{k} \quad , \quad j = 1,\ldots,n \quad , \quad i = 1,\ldots,m \quad . \tag{A.22}$$

One has then,

$$\sum_{i=1}^{m} A_{1i}^{\lambda_1} \ldots A_{ni}^{\lambda_n} = \sum_{i=1}^{m} B_{1i}^{\mu_1} \ldots B_{ni}^{\mu_n} \leq$$

$$\leq \left(\sum_{i=1}^{m} B_{1i}\right)^{\mu_1} \ldots \left(\sum_{i=1}^{m} B_{ni}\right)^{\mu_n} = \left(\sum_{i=1}^{m} A_{1i}^{k}\right)^{\frac{\lambda_1}{k}} \ldots \left(\sum_{i=1}^{m} A_{ni}^{k}\right)^{\frac{\lambda_n}{k}} \tag{A.23}$$

where use has been made of the inequality (A.14) for the $\{B_{ij}\}$ and $\{\mu_i\}$, which has already been proved for the case when the $\{\mu_i\}$ satisfy (A.21).

From (A.23) one can write

$$\frac{\sum_{i=1}^{m} A_{1i}^{\lambda_1}\cdots A_{ni}^{\lambda_n}}{\left(\sum_{i=1}^{m} A_{1i}\right)^{\lambda_1}\cdots\left(\sum_{i=1}^{m} A_{ni}\right)^{\lambda_n}} \leq \left[\frac{\left(\sum_{i=1}^{m} A_{1i}^{k}\right)^{\frac{1}{k}}}{\sum_{i=1}^{m} A_{1i}}\right]^{\lambda_1}\cdots\left[\frac{\left(\sum_{i=1}^{m} A_{ni}^{k}\right)^{\frac{1}{k}}}{\sum_{i=1}^{m} A_{ni}}\right]^{\lambda_n} . \tag{A.24}$$

Now, for any $k > 1$ and any sequence $a = (a_1,\ldots,a_m)$, $a_j > 0 \;\; \forall j$, the elementary inequality

$$\frac{\sum_{j=1}^{m} a_j}{\left(\sum_{j=1}^{m} a_j^k\right)^{\frac{1}{k}}} = \sum_{j=1}^{m}\left[\frac{a_j}{\left(\sum_{j=1}^{m} a_j^k\right)^{\frac{1}{k}}}\right] = \sum_{j=1}^{m}\left[\frac{a_j^k}{\left(\sum_{j=1}^{m} a_j^k\right)}\right]^{\frac{1}{k}} \geq$$

$$\geq \sum_{j=1}^{m}\frac{a_j^k}{\left(\sum_{j=1}^{m} a_j^k\right)} = 1 \tag{A.25}$$

holds, namely

$$\frac{\left(\sum_{i=1}^{m} a_i^k\right)^{\frac{1}{k}}}{\sum_{i=1}^{m} a_i} \leq 1 \quad . \tag{A.26}$$

Then the right-hand side of (A.24) is further ≤ 1, and (A.14) is proved in general.

References

— Chapter 1

• E.T. Whittaker, "Analytical Dynamics", Cambridge University Press, Cambridge, U.K., 1937.

• H. Goldstein, "Classical Mechanics", Addison-Wesley, Cambridge, Mass., 1950.

• R. Abraham and J. Marsden, "Foundations of Mechanics", W.A. Benjamin, New York, 1978.

• V.I. Arnold, "Mathematical Methods of Classical Mechanics", Springer Verlag, Heidelberg, 1978.

• H. Poincaré, "Les Méthodes Nouvelles de la Mécanique Céleste", Gauthier-Villars, Paris, 1899.

• G.D. Birkhoff, "Nouvelles Récherches sur les Systèmes Dynamiques", Mem. Pont. Acad. Sci. Novi Lyncaei 1, 85 (1935).

• J.W. Milnor, "Topology from the Differentiable Viewpoint", The University Press of Virginia, Charlottesville, 1972.

• S. Yorna, ed., "Topics in Nonlinear Dynamics", Amer. Inst. Phys. Conf. Proc. Vol. 46, New York, 1978.

• G. Benedek, H. Bilz and R. Zeyher, "Statics and Dynamics of Nonlinear Systems", Springer Verlag Series in Solid State Sciences, Vol. 47, Springer Verlag, Berlin, 1984.

• J. Moser, "Stable and Random Motion in Dynamical Systems", Princeton University Press, Ann. Math. Studies, Vol. 77, Princeton, 1973.

- R.H.G. Helleman, "Self-generated Chaotic Behaviour in Nonlinear Mechanics", in "Fundamental Problems in Statistical Mechanics", Vol. 5, E.G.D. Cohen, ed., North-Holland, Amsterdam, 1980.
- A.J. Lichtenberg and A.M. Liebermann, "Regular and Stochastic Motion", Springer Verlag, Heidelberg, 1982.
- A.N. Kolmogorov, "On Conservation of Conditionally Periodic Motions for a Small Change in Hamilton's Function", Dokl. Akad. Nauk. USSR 98, 525 (1954).
- V.I. Arnold, "Small Divisors: Proof of a Theorem of A.N. Kolmogorov on the Preservation of Conditionally Periodic Motions under a Small Perturbation of the Hamiltonian", Usp. Mat. Nauk. USSR 18(5), 13 (1963).
- V.I. Arnold, "Small Divisor Problems in Classical and Celestial Mechanics", Usp. Mat. Nauk. USSR 18(6), 91 (1963).
- J. Moser, "Convergent Series Expansion of Quasi-Periodic Motions", Math. Ann. 169, 163 (1967).
- M. Hénon and C. Heiles, "The Applicability of the Third Integral of the Motion: Some Numerical Results", Astron. J. 69, 73 (1964).
- J. Ford, "The Statistical Mechanics of Classical Analytic Dynamics", in "Fundamental Problems in Statistical Mechanics", Vol. 3, E.G.D. Cohen, ed., North-Holland, Amsterdam, 1975.
- R.H.G. Helleman and G. Ioos, eds., "Chaotic Behaviour in Deterministic Systems" (Les Houches Summer School), North-Holland, Amsterdam, 1983.
- V.I. Arnold and A. Avez, "Ergodic Problems of Classical Mechanics", W.A. Benjamin, New York, 1974.
- H. Haken, ed., "Chaos and Order in Nature", Springer Verlag, Berlin, 1981.
- R.S. Shaw, "Strange Attractors, Chaotic Behaviour and Information Flow", Z. Naturfortschr. 36A, 80 (1981).
- E. Artin, "Ein mechanisches System mit quasiergodischen Bahnen", Abh. Math. Sem. Univ. Hamburg 3, 170 (1924).
- P. Billingsley, "Ergodic Theory and Information", J. Wiley, New York, 1965.

- W.H. Gottschalk and G.A. Hedlund, "Topological Dynamics", Amer. Math. Soc. Colloq. Publ., Vol. 36, Providence, 1955.
- Ya. G. Sinai, "Introduction to Ergodic Theory", Princeton University Press, Math. Notes; Princeton, 1976.
- D.V. Anosov, "Roughness of Geodesic Flows on Compact Riemannian Manifolds of Negative Curvature", Dokl. Akad. Nauk. USSR 145, 707 (1962).
- D.V. Anosov, "Ergodic Properties of Geodesic Flows on Closed Riemannian Manifolds with Negative Curvature", Dokl. Akad. Nauk. USSR 151, 1250 (1963).
- M.W. Hirsch and S. Smale, "Differential Equations, Dynamical Systems and Linear Algebra", Academic Press, New York, 1965.
- L. Markus, "Lectures in Differential Dynamics", Amer. Math. Soc. Reg. Conf. Series, Vol. 3, Providence, 1971.
- Z. Nitecki, "Differentiable Mechanics", M.I.T. Press, Cambridge, Mass., 1971.
- S. Smale, "Differentiable Dynamical Systems", Bull. Amer. Math. Soc. 73, 747 (1967).
- E. Hopf, "Abzweigungen einer periodischen Lösung von einer stationären Lösung eines Differentialgleichunssystem", Math. Naturwiss. Klasse, Sächs. Akad. der Wissensch., Leipzig 94, 1 (1942).
- J. Guckenheimer, "Bifurcation and Catastrophe", in "Dynamical Systems", M. Peixoto, ed., Academic Press, New York, 1973.
- D. Ruelle and F. Takens, "On the Nature of Turbulence", Commun. Math. Phys. 20, 167 (1971).
- D.H. Sattinger, "Topics in Stability and Bifurcation Theory", Springer Verlag, Berlin, 1973.
- H. Haken, ed., "Evolution of Order and Chaos", Springer Verlag, Heidelberg, 1982.
- M. Hénon, "A Two-Dimensional Map with a Strange Attractor", Commun. Math. Phys. 50, 69 (1976).
- E. Lorenz, "Deterministic Non-Periodic Flow", J. Atmos. Sci. 20, 130 (1963).

- D. Ruelle, "Strange Attractors", Math. Intelligencer 2, 216 (1980).
- E. Ott, "Strange Attractors and Chaotic Motions of Dynamical Systems", Revs. Mod. Phys. 53, 655 (1981).
- M.J. Feigenbaum, "Quantitative Universality for a Class of Nonlinear Transformations", J. Stat. Phys. 19, 25 (1978).
- M.J. Feigenbaum, "The Universal Properties of Nonlinear Transformations", J. Stat. Phys. 21, 669 (1979).
- T.C. Bountis, "Period Doubling Bifurcations and Universality in Conservative Systems", Physics 3D, 577 (1981).
- O.E. Lanford, "A Computer Assisted Proof of the Feigenbaum Conjectures", Bull. Amer. Math. Soc. 6, 427 (1982).
- P. Collet and J.P. Eckmann, "Iterated Maps on the Interval as Dynamical Systems", Birkhäuser, Boston, 1980.
- A.N. Šarkovskiĭ, "Coexistence of Cycles of a Continuous Map of a Line into Itself", Ukr. Mat. Z. 16, 61 (1964).
- L. Block, J. Guckenheimer, M. Misiurewicz and L.S. Young, "Periodic Points and Topological Entropy of One-Dimensional Maps", in "Global Theory of Dynamical Systems", Z. Nitecki, ed., Springer Verlag Lecture Notes in Mathematics, Vol. 819, Springer Verlag, Berlin, 1979.
- S. Katsura and W. Fukuda, "Exactly Solvable Models Showing Chaotic Behaviour", Physics 130A, 597 (1985).

— Chapter 2

- J.W. Gibbs, "Elementary Principles in Statistical Mechanics", Dover Publ., New York, 1960.
- P. and T. Ehrenfest, "The Conceptual Foundations of the Statistical Approach to Mechanics", Cornell University Press, Ithaca, 1959.
- A. Ya. Khinchin, "Mathematical Foundations of Statistical Mechanics", Dover Publ., New York, 1949.
- O. Penrose, "Foundations of Statistical Mechanics", Pergamon Press, Oxford, 1970.
- P. Glansdorff and I. Prigogine, "Structure, Stability, and Fluctuations", J. Wiley, New York, 1971.

• J. Lebowitz and O. Penrose, "Modern Ergodic Theory", Phys. Today 23(2) (1973).

• Ya. G. Sinai, "Probabilistic Ideas in Ergodic Theory", Transl. Math. Soc. Series 2, 31, 62 (1963).

• A.N. Kolmogorov, "On the Entropy per Time Unit as a Metric Invariant of Automorphisms", Dokl. Akad. Nauk. USSR, 124, 754 (1959).

• Ya. G. Sinai, "Gibbs Measures in Ergodic Theory", Usp. Mat. Nauk. USSR 27(4), 21 (1972).

• D. Ruelle, "Thermodynamic Formalism", Addison-Wesley, Reading, 1978.

• Ya. I. Dinaburg, "The Relationship Between Various Entropic Characteristics of Dynamical Systems", Izv. Akad. Nauk. USSR, Ser. Mat. 35(2), 324 (1971).

— Chapter 3

• A. Blanc-Lapierre and R. Fortet, "Théorie des Fonctions Aléatoires", Masson, Paris, 1953.

• N. Bourbaki, "Élements de Mathématique, Livre V: Espaces Vectoriels Topologiques", Hermann, Paris, 1964.

• I.M. Gelfand and N. Ya. Vilenkin, "Generalized Functions, Volume 4: Applications of Harmonic Analysis", Academic Press, New York, 1964.

• W. Feller, "An Introduction to Probability Theory and its Applications", J. Wiley, New York, 1959.

• A. Papoulis, "Probability, Random Variables and Stochastic Processes", McGraw Hill, New York, 1965.

• L. Schwartz, "Théorie des Distributions", Hermann, Paris, 1959.

• A.M. Yaglom and I.M. Yaglom, "Probabilité et Information", Dunod, Paris, 1959.

• A.M. Yaglom, "An Introduction to the Theory of Stationary Random Functions", Prentice-Hall, Englewood Cliffs, 1962.

— Chapter 4

- K. Ito, "Random Processes", Inostrannoj Literatury, Moscow, 1963.
- J.L. Doob, "Stochastic Processes", J. Wiley, New York, 1953.
- E.B. Dynkin, "Théorie des Processus de Markov", Dunod, Paris, 1963.
- N.U. Prabhu, "Stochastic Processes", MacMillan Co., New York, 1965.
- A. Scheerer, "Probability on Discrete Sample Spaces", International Textbook Co., Scranton, 1969.
- S. Grossmann and S. Thomae, "Invariant Distributions and Stationary Correlation Functions of One-Dimensional Discrete Processes", Z. Naturforsch. 32A, 1353 (1977).
- M. Metropolis, M.L. Stein and P.R. Stein, "On Finite Limit Sets for Transformations of the Unit Interval", J. Comb. Theory 15A, 25 (1973).
- P.R. Halmos, "Measure Theory", D. van Nostrand, Princeton, 1950.
- N. Packard, J.P. Crutchfield, J.D. Farmer and R.S. Shaw, "Geometry from a Time Series", Phys. Rev. Lett. 45, 712 (1980).
- H. De Long, "Randomness and Gödel Incompleteness Theorem", Addison-Wesley, Reading, 1970.
- C. Berge, "Théorie des Graphes et ses Applications", Dunod, Paris, 1963.

— Chapter 5

- C.E. Shannon and W. Weaver, "The Mathematical Theory of Information", University of Illinois Press, Urbana, 1949.
- A. Ya. Khinchin, "Mathematical Foundations of Information Theory", Dover Publ., New York, 1958.
- I.M. Gelfand, A.N. Kolmogorov and A.M. Yaglom, "On a General Definition of Amount of Information", Dokl. Akad. Nauk. USSR 111(4), 745 (1956).

• D.A. Fadiev, "On the Notion of Entropy of a Finite Probability Space", Usp. Mat. Nauk. USSR 11(1) 227 (1956).

• J.D. Farmer, "Information Dimension and Probabilistic Structure of Chaos", Z. Naturforsch 37A, 1304 (1982).

• B. McMillan, "The Basic Theorems of Information Theory", Ann. Math. Stat. 24, 196 (1953).

• M. Rosenblatt-Roth, "On the Entropy of Stochastic Processes", Dokl. Akad. Nauk. USSR 112(1), 1957.

— Appendix

• G.H. Hardy, D.E. Littlewood and G. Polya, "Inequalities", Cambridge University Press, Cambridge, 1952.

Subject Index